THE BEGINNER'S GUIDE
to growing
HEIRLOOM VEGETABLES

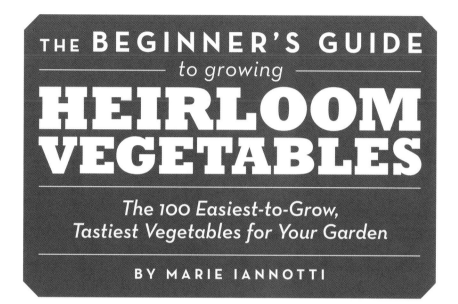

THE **BEGINNER'S GUIDE**

to growing

HEIRLOOM VEGETABLES

The 100 Easiest-to-Grow,
Tastiest Vegetables for Your Garden

BY MARIE IANNOTTI

TIMBER PRESS
PORTLAND | LONDON

Frontispiece: The colors, shapes, and flavors of heirloom vegetables are just a hint of the diversity available.

Photography credits appear on page 250.

Published in 2011 by Timber Press, Inc.

The Haseltine Building
133 S.W. Second Avenue, Suite 450
Portland, Oregon 97204-3527
timberpress.com

2 The Quadrant
135 Salusbury Road
London NW6 6RJ
timberpress.co.uk

Printed in China

Library of Congress Cataloging-in-Publication Data

Iannotti, Marie.
  The beginner's guide to growing heirloom vegetables: the 100 easiest-to-grow, tastiest vegetables for your garden/Marie Iannotti.—1st ed.
      p. cm.
  Includes bibliographical references and index.
  ISBN 978-1-60469-188-7
  1. Vegetable gardening. 2. Vegetables—Heirloom varieties. I. Title.
  SB321.I36 2012
  635–dc23
                                                              2011018441

A catalog record for this book is also available from the British Library.

To my mother, who taught me to love vegetables, and my father, who taught me to grow them.

CONTENTS

Acknowledgments

I am indebted to all the wonderful writers and seed savers who have made sure heirloom vegetables are still available to us. You have whetted my appetite to learn and grow more and made my vegetable garden such an adventure. Special thanks to the folks at Baker Creek Heirloom Seeds and Seed Savers Exchange for helping me with questions, and to Anna de Cordova, Susan MacAvery, and all the extraordinary volunteers at Locust Grove, the Samuel F. B. Morse Historic Site, for the amazing variety I was able to sample during my years with you. My appreciation also goes to all the educators in the cooperative extension system, for their generosity in sharing their wealth of knowledge.

I am grateful for the continual encouragement from my friends, Marge Bonner, who grew many of the plants along with me and served as my backup and taste tester, and Cheryl Alloway, who let me vent and whose advice has made my garden better. Thank you to my brother, John Iannotti, who never turned down my tomatoes and beans.

I would never have muddled through without the help of Juree Sondker and all the folks at Timber Press, who guided me through, and Lisa Theobald, who pulled it all together.

Finally, huge thanks to my husband, Michael Juzwak, for his unflagging support and patience while I took over every inch of the yard.

HEIRLOOM VEGETABLES AND WHY WE PRIZE THEM

GROWING WONDERFULLY SATISFYING FOOD is the main point of vegetable gardening, and the world of heirloom vegetables is so excessively abundant, it can leave you feeling giddy. A vegetable garden in full production reminds me of the dessert cart in a fine restaurant. Where to focus, what to choose? Must it be only one? Homegrown heirloom vegetables can be so beautiful and delicious that it seems you could simply inhale them, and many vegetables never make it all the way from the garden to the kitchen. Asparagus, peas, and tomatoes, to name a few, make quick and delicious snacks, with no cooking required.

I have an obsessive compulsion for heirloom vegetables, and I blame my condition on my dad. After gardening all his life, he was becoming disillusioned with his efforts. He had the audacity to complain that our tomatoes were not as good as those in his memory. Although I started my gardening career as a child laborer in the backyard plot, I had developed a certain pride and competitiveness and was not about to be told my tomatoes were subpar. Then I read somewhere that perhaps we were not growing the right types of tomatoes. What I learned is that if you want the flavor of the tomatoes from your childhood, you should grow those same tomatoes. I am certain many heirloom converts were made this way. Start with a 'Brandywine' tomato and you never know where it will lead.

Heirloom beets with such intriguing names as 'Crosby's Egyptian' and 'Detroit Dark Red' delight all the senses with their saturated colors, dense flesh, and full flavor.

11

For me, there is so much more to heirloom vegetables than childhood memories of sun-sweetened tomatoes. While writing this book, I experienced the daily joy of walking out to my garden and tasting delights from a parade of cultures. I like to think of it as the summer I grew the world and ate it. My garden included many more than the 100 vegetables featured here; I would sample each, ruminate a bit, close my eyes like I was sipping fine wine, and try to come up with descriptions that conveyed more than just, "Wow, that's good!" My hedonistic gluttony and being surrounded by the psychedelic array of colors and scents often made me forget the task at hand. The qualities and idiosyncrasies of these vegetables can be dense and sweet; rich and spicy; striped, speckled, and splotched; and so unendingly unique at every turn—a menu with limitless possibilities.

As I chose vegetables to feature in these pages, I tried not to play favorites, but some heirlooms are simply brazen. When the sun shines through a 'Golden Sweet' sugar pea, there is no time to grab your wok; the temptation to pop one in your mouth is too strong to ignore. The sun's warmth is enough to release its nutty, honey crunch with no loss of luminescence. In a league of its own is the rat's tail radish (Raphanus sativus 'Caudatus'). Try convincing your friends that they will enjoy eating a vegetable named for a rodent's tail. It had better be very good—very, very good. And it is. I watched my friend Marge (my backup gardener, who was persuaded to grow more vegetables in her own garden than her sanity cautioned) delight a group of master gardeners who tasted her rat's tail radishes. Everyone asked for seeds, which is the foundation of heirloom vegetable gardening: vegetables so irresistible that their seeds are handed from gardener to gardener.

Not every vegetable I grew turned out to be a winner. My former fascination with 'Strawberry' popcorn, for example, must have been with the novelty of its tiny, colorful ears, not its flavor. The ears are only a couple of inches long, and although they are a lovely garnet red, they pop up white and taste unimpressively bland. All was not lost, however, because the ears made a great table centerpiece.

I also had some disappointments. 'Moon and Stars' watermelon is visually captivating, but the melons take forever to mature, even with black plastic laid underneath them to warm the soil. Thank goodness for the heirloom 'Blacktail Mountain', with its shiny, bowling ball–sized fruits that obliged my shorter growing season with crunchy, juicy, sweetly aromatic slices that I could enjoy during the hot, hazy days of August.

For sheer abundance, I am always delighted to grow beans. Every novice heirloom gardener should grow all kinds of beans. Freshly picked beans have a pungency that quickly dissipates on the grocer's shelf, and a glance at the heirloom options will leave you wondering how you ever settled for a mundane green bean. Beans do not require much more than sun and water. A few seeds will quickly turn into twining poles of snappy 'Lazy Housewife', densely rich 'Romano', or fanciful 'Chinese Red Noodle' beans. And we cannot forget shelling beans, like 'Christmas Lima' and others such as 'Speckled Cranberry' and 'Tiger's Eye', which are so colorful and shiny you would think they were gems to be made into jewelry.

What Is an Heirloom Vegetable?

If your idea of an heirloom is your aunt Sally's cameo brooch, rather than a jar of her dried beans, remember that an heirloom can be anything of value that is passed down through generations, be it jewelry, baseball cards, your first grade report card, or a humble jar of beans. Those beans could be the last existing seeds of that variety, making them all the more valuable in general. At the least, the seeds tell us that someone thought that this particular variety of bean was so good, it was worthy of saving.

Because of the evolving or haphazard nature of heirloom varieties, a strict definition of an heirloom vegetable does not really exist. Most sources, how-ever, are in agreement that heirlooms must meet certain requirements.

- They must be open pollinated. The seeds produced by open pollination among plants of the same variety will grow into plants with the same char-acteristics as the parents. This is also referred to as "true to type." By con-trast, most modern seeds are hybrids, crosses between plants that will not produce seed that grows true-to-type plants.
- They must be more than 50 years old. This is an arbitrary qualification, but it has worked fairly well and allows more and more time-tested varieties to

All heirlooms are open-pollinated and can be started from seed; you can even save your own seeds for replanting.

fall into the category of heirloom. The heirloom palette is always expanding. An open-pollinated treat stumbled upon today can become an heirloom to future generations of gardeners. Some young heirlooms are allowed to sneak in because of their pedigree, being stabilized crosses of two bona fide heirloom plants.

- They must be storied or historic. Some purists think that only seeds handed down within families are true heirlooms, as opposed to those handed down through groups or communities. Many of the stories behind heirlooms may be folklore, as few people thought to document the history of their seeds, but the point is becoming moot as heirloom seed catalogs blur the distinction between true heirlooms and the others.

With the plethora of seed sources available today, it can be difficult to imagine a time when gardeners could not simply order new seeds from a catalog every spring. But not so long ago, if you wanted a garden, you either saved your own seeds or traded seeds with other gardeners. When we save seeds, we tend to be selective, so the seeds handed down from generation to generation are often among the best vegetable varieties ever grown. More than heirlooms, these are prized champions.

As such, it might be a disservice to label vegetables as merely heirlooms. They are not dusty, fragile, or exclusive, and they are not part of some trendy movement or quirky lifestyle. Although they certainly have earned their place in history, heirlooms are still vegetables intended for the table. Their true splendor comes from being too scrumptious to forget, so we continue to grow and eat them for generations. Contrast this with modern hybrids that were developed for commercial farming, varieties bred to ship without bruising, to ripen on trucks, and be uniform in size and shape. During the trip from your backyard garden to your kitchen, bruising will not be an issue, and you can harvest each unique vegetable at the peak of its ripeness.

Growing and eating heirloom vegetables gives you a direct connection with gardeners and cooks from centuries past, not to mention gardeners from your own past. I still remember my father and his uncle debating the merits of 'Roma' and 'San Marzano' paste tomatoes over 'Big Beef' beefsteak tomatoes as we stood in uncle Vito's greenhouse at the start of the growing season. Why, I wondered, would anyone need that many tomato plants? But like so many gardeners who grew up without the convenience of tomatoes being available year-round, my Dad's uncle was taking no chances that his favorites might not be on the table to enjoy that summer.

Of course, heirlooms are not perfect—they are plants, after all. Not all varieties are suited to all climates and growing conditions, but if you select the hardiest seed and continue to grow it year after year, it should eventually acclimate to your region. Heirlooms could not have lasted as viable garden plants if they were as temperamental as hybrid zealots would have you believe. To be fair to hybrids, many good varieties are available, and some offer improved disease resistance and vigor. But hybrids are one-size-fits-all, and they lack both the regionalism of heirlooms and their broad and distinct flavors. Every garden has room for a little of both.

Heirloom vegetable plants can be compact and well-behaved, or they can be space hogs and ugly ducklings. For every sugary 'Minnesota Midget' melon that is just the right size to split between two people, there are sprawling plants such as the crunchy 'Mammoth Red Rock' cabbage and 'Giant Musselburgh' leeks that like some elbow room in the garden.

And this brings us to the serious side of heirlooms: their genetic uniqueness. As we grow less and less variety, some heirlooms are disappearing from seed catalogs. Over the years, heirlooms have adapted to the climates of the regions where they are grown and have developed natural resistance to pests and diseases. Plant breeders and other researchers rely on the genetic diversity of these open-pollinated plants to develop new varieties. There would be no hybrids without heirlooms.

Will we ever tire of the first sun-ripened tomato of summer? Will we ever

(Above) The licorice flavor and aroma of Italian finocchino (fennel) can be grown to perfection in your garden. Growing heirloom vegetables is like taking a trip around the world without ever leaving your own backyard. (Below) Hundreds of crunchy, colorful heirloom lettuce varieties can be tucked throughout your garden. They are fast-growing, beautiful, and always taste best when fresh.

give up the quest to find the perfect tomato of our childhood? Perhaps the most compelling reason for sampling an array of heirloom vegetables in your garden is simply to become aware of the choices that are available to you. You can grow the whole world in your garden. When you have the opportunity to grow the very best, at no extra cost or effort, why settle for anything less?

Getting Started Growing Heirloom Vegetables

How do you judge a vegetable worthy of the limited space in your vegetable garden? What are the characteristics that make a star? Your first criteria should always be taste. That was my main criteria for the 100 vegetables profiled in these pages. Each has something unique to bring to the table and none is terribly fussy or demanding in the garden—another stellar feature.

Everyone has particular favorites, so grow what you like. If you and your finicky 5-year-old both have a sweet tooth, you can plant sugary 'Golden' beets or some plump 'Paris Market' carrots. If you do not favor eggplant, even a beautiful 'Rosa Bianca' will go to waste. But sight is seductive, so flip through this book and consider the temptations posed by a jewel-toned tomato, a speckled melon, or the glossy sheen of a purple tomatillo, half hidden in its papery sheath.

It is impossible for researchers to pinpoint the exact origins of specific vegetables—where they first grew wild and where they were first cultivated. The fact that they traveled the world and became assimilated in different cultures is far more interesting. The tomato, for example, is not native to Italy, but just look at what Italians have done with it.

As with wine grapes, a vegetable's quality, size, and taste can vary, depending on soil, sun, and other growing conditions. Grape vines produce the finest wine grapes in soils that are a bit lean (not too rich in organic matter) and loose enough to allow the roots to grow deeply, sustaining the plants without much supplemental water. The French refer to this set of growing conditions as terroir. Although most seasonal vegetables do not have time to send down deep roots, they are still affected by environmental conditions. Flavors, intensities, and reliability of heirlooms are somewhat contingent on the region in which they are grown. For instance, tomatoes grown in regions with a lot of rain and too little sun during the growing season can have a lower sugar content, resulting in less sweetness, even when they are fully ripened on the vine. Hot peppers, on the other hand, become more potent when grown in lean, drier soil.

Vegetable gardens are a tantalizing mix of flavors, fragrance, and spectacle. An intrinsic connection exists between vegetable gardener and cook; the primal scent of chestnut squash or the dazzle of melons in yellow, red, green, and glowing orange can inspire and excite. Treat your garden as a vegetable sampler and invite others to join you in tasting them. There is no easier way to start a conversation with friends, old or new, than to take them on a tasting tour of your vegetable garden. Heirloom vegetables are stories waiting to be told, impressions waiting to be made.

What will you discover in your garden?

100 FAVORITE HEIRLOOM VEGETABLES

Although I found it difficult to limit my selections to 100 vegetables, doing so was one of the most delectable challenges I have ever faced. Should I go with the buttery and beautiful 'Forellenschluss' lettuce or the crispy work horse 'Marvel of Four Seasons'? Should I favor the cool heat of 'White Icicle' radish or leave room for delicate 'Applegreen' eggplant? I grow lots of spicy 'Friggitello' peppers to put up for winter, but I cannot live without the tangy 'Long Red Florence' onions.

Because I had to limit my choices to 100, I chose varieties that offer flavors you will not find elsewhere and hardy plants that show off their wares. As I ate my way through these vegetables, trying to come up with evocative and accurate words to communicate earthy root vegetable and succulent melon, I came to appreciate that these are joyous foods that say, "Sit down. Grab a fork. Where have you been?"

'German Butterball' potatoes really do melt in your mouth. 'Bistro' mache stays sweet and tender despite the heat. 'Rampicante Tromboncino' zucchini holds its delicate, firm texture no matter how overcooked. In the end, I chose the varieties I return to year after year, hoping you will enjoy them as much as my dinner guests and I do.

MARIE'S TOP PICKS

To help you find the plants and flavors that suit your garden and your palate, I offer ten particular categories of heirloom vegetables that have distinguished themselves above the rest. Each top pick is described in detail.

AROMATIC

Scent plays a big role in flavor, and these ambrosial vegetables will entice you to the table, before you even see them. *Fennel 'Zefa Fino', Garlic 'Spanish Roja', Melon 'Minnesota Midget'*

BEAUTIFUL

We eat with our eyes first. Never underestimate the power of a stunning vegetable, beautifully prepared. *Bean 'Dragon's Tongue', Cauliflower 'Di Sicilia Violetto', Pea 'Golden Sweet'*

CLASSIC

Certain vegetables have been breakout stars since gardeners first started paying attention to heirlooms. These classics have stood the test of time and are still best sellers today. *Bean 'Kentucky Wonder', Beet 'Chioggia', Spinach 'Bloomsdale Long Standing'*

COLORFUL

Beet red and lemon yellow are not always the norm. One of the most enjoyable aspects of heirloom gardening is the variety offered, and many heirlooms can surprise you with their colorfulness. *Eggplant 'Turkish Orange', Squash 'Sweet Dumpling', Swiss chard 'Rainbow Chard'*

LONG SEASON

These vegetables chug along for the entire growing season, allowing you to plan meals around them or harvest them in a pinch. Rather than ripening all at once, these winners are slow and steady growers. *Amaranth 'Red Leaf', Lacinato kale, Okra 'Burgundy'*

PROLIFIC

These heirlooms will provide an abundance of vegetables to enjoy fresh, preserve, or share. *Collards 'Vates', Cucumber 'Crystal Apple'*

SPICY

Spicy vegetables can add a touch of heat and contribute their complex flavor to any meal. *Pepper 'Fatali', Radish 'Round Black Spanish'*

SWEET

Who says vegetables have to be savory? Several are sweetly luscious. *Beet 'Golden', Carrot 'Paris Market', Melon 'Boule d'Or'*

UNUSUAL

Let's face it; sometimes you cannot judge a book by its cover. Delicious as they are, these vegetables are just plain odd looking. *Asparagus pea, Rat's tail radish, Zucchini 'Rampicante Tromboncino'*

VERSATILE

When space and time are an issue, you can grow vegetables that offer you flexibility in the kitchen. These heirlooms make perfect additions, no matter what you are preparing. *Radish 'Red Meat', Zucchini 'Ronde de Nice'*

ARTICHOKES AND JERUSALEM ARTICHOKES

Looking for a savory indulgence? Try growing the dense, earthy flavors and creamy textures of these members of the sunflower family. Globe artichoke and Jerusalem artichoke might not seem to have much in common other than their names, but both are members of the family Asteraceae, a huge, raucous family that includes some much loved flowers such as Shasta daisies and blanket flowers, as well as thistles, weeds, and many salad greens. All the flowers in this family actually comprise clusters of hundreds of tiny flowerets, such as the fuzzy choke that covers an artichoke's tender heart.

At some point in history, both vegetables were considered weeds. Jerusalem artichokes will easily spread on their own, but globe artichokes are more particular about where they grow.

Globe artichokes and Jerusalem artichokes have one more common feature important for gardeners to know: both require a long, warm growing season. So get started early.

'Purple of Romagna' artichoke is such a tender little thistle, it blushes when it is ready to be picked. It is both stunning and delicately delicious. These towering architectural plants, named for the region in southeastern Italy where they were popularized, have gray-green leaves and supple, baby-sized buds, suffused with purple.

Exposure: Full sun

Ideal soil temperature: 70–80°F (21–27°C)

Planting depth: 1/4 in. (0.6 cm)

Days to germination: 10–14 days

Spacing: 4–8 ft. (1.2–2.4 m) when grown as a perennial, 3–4 ft. (0.9–1.2 m) when grown as an annual

Days to maturity: 90–110 days

Artichoke 'Purple of Romagna'

Globe artichoke
Cynara scolymus

Flavor 'Purple of Romagna' has a heartier flavor than its green artichoke cousins, with a deep, earthy, nutty-olive taste. Almost the entire leaf is edible, with very little choke. When eaten young and tender, they require virtually no cooking, but a quick sauté with olive oil or butter and a splash of lemon juice will bring out their vibrant flavor.

Growing notes Artichokes are perennials that tend to bloom in their second year. If you garden where winter temperatures dip below freezing, you can grow them as annuals by tricking the plants into thinking they have already gone through a winter. To do this, start them indoors in early spring and move them to a cold frame or protected area, where they will chill at about 50°F (10°C) but will not freeze. Keep them in the cold frame for 3–6 weeks, and then move them into the garden when the soil warms. Gardeners in Zones 8 and warmer can start plants in the fall to harvest the following spring. You can start seed indoors and move them out as seedlings, or direct seed them outdoors. You can also start plants from dormant roots called stumps.

Prepare the soil by adding plenty of rich organic matter, especially if you will be growing artichokes as perennials. Keep the plants well watered. Ideal growing conditions include rich soil, mild winters, and cool springs. Dry soil can produce tough artichokes that open prematurely.

If you garden where winter temperatures reach freezing or lower, your plants will need some protection. Cut back the plant to about 10 in. (25 cm), mulch heavily, and cover the plant with a bucket or box. Alternatively, you can overwinter your plants in containers indoors. Move the plant to a cool, dark spot and give it occasional water. Move it back outdoors after all danger of frost has passed.

How to harvest A mature 'Purple of Romagna' sets many buds during a short harvest season. The central bud will mature first, followed by side shoots. Cut the buds and about 2 in. (5 cm) of stalk before the bracts begin to open. Buds will feel dense and heavy when ripe. Uncut buds will open into thistle flowers.

Tip Do not let your artichoke flowers go to seed, or seed will self-sow and weedy thistles will spread throughout your garden; artichoke seed rarely grows true to type.

Others to try 'Romanesco' is a large purple choke that matures midsummer and is a

good choice for shorter seasons. 'Violetta di Chioggia' is a deep purple, spineless variety. 'Violetta Precoce' is an elongated choke and a good choice for mild climates.

(Opposite) Both the artichoke plants and the edible flower buds offer beautiful architectural interest in the garden and are a delight on the plate.
(Above) To keep Jerusalem artichokes from spreading in the garden, harvest the tubers while they are small and tender.

Exposure:
Full sun to partial shade

Ideal soil temperature:
45–50°F (7–10°C)

Planting depth:
3–5 in. (7.5–12.5 cm)

Days to germination:
10–17 days

Spacing:
24–26 in. (61–66 cm)

Days to maturity:
120–140 days

This native American shows the initiative and gumption of its colonial roots. Jerusalem artichoke is neither from Jerusalem, nor is it an artichoke. It is a species of sunflower, but we eat the knobby, unpretentious tuber. It fell out of favor for years but has resurfaced with our renewed interest in native plants and perennial vegetables and is often marketed today as a sunchoke. Until recently, you could not find one with a varietal name. Many varieties offered today are simply named after the place in which they were found, and some may actually be the same plants but with different names. They are all probably heirlooms, however, because Jerusalem artichokes are grown from tubers, not from seed.

Artichoke, Jerusalem
Sunchoke
Helianthus tuberosus

Flavor Jerusalem artichokes have a nutty flavor similar to that of globe artichokes, with a texture more reminiscent of potatoes. It is a versatile vegetable that can be sliced raw into salads to offer a crunchy texture or cooked in any number of ways, even whipped into a smooth puree. The raw tubers are juicy—similar to a cross between water chestnuts and celery, with a fresh sweetness. They also make great pickles.

Growing notes Start plants from whole tubers or pieces of tubers with at least two or three buds. Unlike potatoes, artichoke pieces should not be allowed to dry out before planting. Jerusalem artichokes grow larger in Zones 8 and warmer. These zones offer the long growing season that they favor.

Plant pieces in early spring, after the ground has dried to a crumbly texture and all danger of frost has past. Although they need a long growing season to form tubers, do not rush to plant them; the plants need warm weather to start growing.

They should not require supplemental feeding if planted in rich soil, but keep them well watered. Although they can grow quite tall, to 6-8 ft. (2-2.4 m) or taller, you can pinch the plants' tips to maintain their height at about 18-20 in. (45.5-51 cm), which will improve the size of the tubers.

Few pests are known to bother Jerusalem artichokes, other than competitive weeds.

How to harvest Wait until a frost has touched the foliage before digging up the tubers. The chill will sweeten them. Cut back the tops and fork over the soil to loosen the tubers. You will need to do some sifting, because Jerusalem artichokes can be small.

Tip Jerusalem artichokes are aggressive growers. Growing them in some type of contained or raised bed will help prevent them from spreading.

Others to try 'Mammoth French White' is a late bloomer with large, knobby, white tubers. 'Red Fuseau' has smooth skin and maroon tubers and is heat and drought tolerant. A good choice for poorer soils is 'Stampede', with large, knobby, white tubers; it is prolific and flowers early.

ASPARAGUS

Some of the most heralded spring sights are the braided tips of asparagus poking their heads above ground. The first spring crop is always the sweetest and most tender. Snap off the spears at their base; they are delicate enough to be eaten without cooking, with a lively, fresh flavor. Asparagus season is fleeting, but furious. Each plant will produce spears for 6 to 8 weeks, growing faster and faster as the days warm up. A spear can shoot up 10 in. (25 cm) in a single day.

Asparagus has been cultivated for more than 2000 years. The Romans were especially fond of it, leaving us detailed growing instructions. A popular expression of their day, *Citius quam asparagi coquentur* (Do it quicker than you can cook asparagus), shows their reverence for and understanding of how best to enjoy asparagus.

'Mary Washington' woos you with her beauty and draws you back with her flavor. The plump, heavy spears are dark green with a hint of purple, with new spears popping up for weeks. Developed for disease resistance, it was introduced to gardeners in 1949 and was the most popular asparagus on the block until all-male varieties were introduced. The boys may produce a few more spears, but they have nothing on Mary's tenderness and flavor.

Asparagus 'Mary Washington'

Asparagus officinalis

Flavor 'Mary Washington' has that full-blown tangy savoriness associated with fresh asparagus. The spears are succulent and tender, retaining their crispness even as they plump up. Their lovely snapping green freshness reminds me of the brisk, clean air of spring. Cook asparagus gently to maintain its flavor. Steaming with lemon or roasting with olive oil will keep the bright green color and full flavor.

Growing notes Most gardeners prefer to start their stands with 1- to 2-year-old crowns—buds with long, moplike roots. Choose crowns with firm, dry roots. Crowns can be planted in early spring, once the soil has warmed to about 50 degrees F. (10 C.). You can start asparagus from seed indoors, about 2–3 months before your last frost date, and then transplant them outdoors after all danger of frost has past. In warm climates, seed can be started in late summer through early autumn, and crowns can be planted midautumn through early spring.

Prepare a trench 12–18 in. (30.5–45.5 cm) wide and 9–12 in. (22.5–30.5 cm) deep, with the soil slightly hilled in the center of the trench. Place the crowns on the hill and spread the roots out evenly. Then cover the roots with 1/2 in. (1 cm) of soil and keep filling in the trench, barely covering the plants as they grow.

For the first two years, asparagus will produce lots of roots and spindly spears. Be patient, because you must wait to harvest most asparagus until after 3 years of growth. Once established, asparagus plants can continue producing for 15 to 20 years.

Because asparagus is a perennial plant, it will need extra fertilizer. For the first 3 years, apply a balanced fertilizer in early spring. Beginning in the fourth year, do not feed the plants until the harvest is complete. Allow the plants to die back naturally in the fall, to act as mulch throughout the winter.

How to harvest It usually takes 3 years for asparagus plants to be strong enough to begin harvesting, but you can lightly harvest some spears in the second year. After about a month of growing, the new spears will begin to be thinner in diameter. Let them grow and continue to feed the plant. By their fourth year in the ground, you will be harvesting throughout the spring.

Harvest spears when they are

Exposure:
Full sun to partial shade

Ideal soil temperature:
65–75°F (18–24°C)

Planting depth:
Seeds, 1/2 in. (1 cm)

Days to germination:
10–15 days

Spacing:
18–24 in. (45.5–61 cm)

Days to maturity:
3 years

'Mary Washington' will keep sprouting for months if you keep harvesting the young spears.

5–8 in. (12.5–20 cm) tall. You can slice them off just under the soil line, but you run the risk of damaging emerging spears. Snapping the spears above the soil line is better and avoids this problem.

Tip Asparagus plants can rise or lift out of the soil over time; adding a few inches of soil over the crowns every few years can help.

Others to try 'Argenteuil' is an early French variety that is purple in color but is blanched (covered in soil to block the sunlight and the creation of chlorophyll) to grow tender white asparagus. 'Conover's Colossal' is fat and prolific. 'Purple Passion' is a tender, sweet purple asparagus. The purple disappears with cooking.

BEETS

Beets are believed to have been cultivated since before recorded history, but no one needs to know that those trendy terrestrial globes on the plate are actually prehistoric. Clean and slice these little diamonds in the rough and you have sparkling, jewel-toned rings with a juicy crispness and one of the highest sugar contents of any vegetable. Beets also pack a big bang of nutrition, especially when you harvest them while they are still small and tender. Grate them fresh, roast them until caramelized, pickle them in marinade, or simply sauté their greens: all are delicious ways to eat beets.

'Bull's Blood' does not form much of a bulb, but the greens are gorgeous and tasty. (Below)

Exposure:
Full sun to partial shade

Ideal soil temperature:
75°F (24°C)

Planting depth:
1/2–1 in. (1–2.5 cm)

Days to germination:
5–10 days

Spacing:
2–4 in. (5–10 cm)

Days to maturity:
55–65 days

'Bull's Blood' is beautiful, with glossy burgundy leaves that intensify as they age. They have been grown as ornamentals, but that seems like a waste, considering how delicious they taste and how easy they are to prepare. These beet greens are sweeter in flavor and firmer in texture than spinach or chard. Even kids will eat them.

Heirloom vegetable names are always descriptive and sometimes downright graphic. Blood beets were named for the thick red juice produced when they are cooked. 'Bull's Blood' beets are such an intense red that they are used to make red food coloring.

Beet 'Bull's Blood'
Beet greens
Beta vulgaris

Flavor 'Bull's Blood' leaves are less earthy flavored than most beet greens. They are almost as tender and sweet as lettuce, with more vitamins. Pick them young and you can use them fresh or slightly wilted in a salad. This beet's root is also savory when roasted, and the leaves can be roasted along with the small and tender bulbs.

Growing notes Direct sow seed in the garden after the soil has dried out and warmed to 50°F (10°C). To maintain a steady harvest throughout the summer, plant beet seeds in succession every 3 weeks until midsummer. In the late summer or early fall, after nighttime temperatures have cooled off to at least 75°F (24°C), you can resume planting seeds for a fall harvest. Plant at least 1 month before the first expected frost date.

Beets do not transplant well and are best planted from seed. Soaking beet seeds overnight will soften their tough coating and speed germination. Beet seeds come in clusters and can be difficult to separate, so some thinning of seedlings is required.

'Bull's Blood' are grown primarily for the leafy tops, so they do not need as much room as beets grown for their bulbs. Thin the seedlings and add them to salads or soups.

Consistent water is the key to keeping beets from becoming woody and keeping the greens fresh. They need at least 1 in. (2.5 cm) of water each week, more during hot, dry weather.

How to harvest Start harvesting when the tops are 2–6 in. (5–15 cm) tall, when they are most tender. Bulbs do not grow large and are ready to harvest when they have reached 1^{1}/2–2 in. (4–5 cm) in diameter. Lift the bulbs by tugging or digging.

Tip Pick the outer leaves as you would leaf lettuce. This will keep the newer inside leaves as sweet and tender as the first harvest.

Others to try 'Early Wonder Tall Top' grows tall, with bright green leaves that stay tender. 'Green Top Bunching' has long, green leaves that can be eaten raw when young and hold their bright color when cooked. 'Mac-Gregor's Favorite' has long, narrow, glossy purple leaves that taste terrific uncooked.

Sweet and striking, 'Chioggia' have made beets popular again.

Exposure:
Full sun to partial shade

Ideal soil temperature:
75°F (24°C)

Planting depth:
1/2–1 in. (1–2.5 cm)

Days to germination:
5–10 days

Spacing:
2–4 in. (5–10 cm)

Days to maturity:
55–65 days

Slice into a 'Chioggia' (pronounced keeohgeeya) beet and you will quickly discover why they are also called the candy stripe beet and bull's eye beet. Although they look like average red beets on the outside, inside they are ringed in alternating shades of brilliant scarlet and white. This Italian heirloom is almost as sweet as candy, but it still has the earthy taste of a great beet.

'Chioggia' is named for a coastal town on an island just south of Venice. The beets are believed to be from a nearby hill town, Bassano, so you might see them listed under that name as well.

CLASSIC

Beet 'Chioggia'
Candy stripe beet, bull's eye beet
Beta vulgaris

Flavor Most beet-lovers agree

that 'Chioggia' beets are a little milder and sweeter than common beets. They tend to have a higher sugar content, a good thing to keep in mind when you are cooking them. They do not "bleed" as much as totally red beets and the stripes often disappear during cooking. The flesh is tender and can be slightly cooked or pickled. The green tops are also edible, with a mild, spicy flavor.

Growing notes Direct sow seeds about 2–4 weeks after the last frost date. To maintain a steady harvest, plant beet seeds in succession every 3 weeks until midsummer. In the late summer or early fall, after nighttime temperatures have cooled off to at least 75°F (24°C), you can resume planting seeds for a fall harvest. Plant at least 1 month before the first expected frost date.

Beets do not transplant well and are best planted from seed. Soaking beet seeds overnight will soften their tough coating and speed germination. Beet seeds come in clusters and can be difficult to separate when planting, so thinning is usually required when the sprouts are a few inches tall; add the thinnings to salads and soups.

Beet roots grow half in the ground and half above ground, so do not plant seeds too deeply and thin the plants to about 2–3 in. (5–7.5 cm) apart. Give the bulbs plenty of room to develop, and the leafy tops will form a canopy to keep the soil moist and cool.

How to harvest Greens can be harvested after they reach 2–6 in. (5–15 cm) tall, while they are still tender. Bulbs are ready to harvest when they have reached 2–3 in. (5–7.5 cm) in diameter. Larger bulbs can become fibrous and tough. Lift the bulbs by tugging or digging.

Tips The beets' stripes often disappear during cooking. If you want to retain the colors, bake them whole and slice them just before serving. Better yet, pick them young and eat them raw.

Not all sliced 'Chioggia' beets display perfect bull's eyes, so do not think you have done something wrong if some beets show more red or more white or some odd combination in between. They still taste delicious.

Others to Try 'Albino' is a creamy white, sweet beet. 'Crapaudine' is an elongated beet with tough skin and sweet flesh. 'Detroit Dark Red' is a standard, tender beet that stores well.

'Golden' beets can be difficult to germinate, but their unique sweetness is worth the extra effort.

Exposure: Full sun to partial shade

Ideal soil temperature: 75°F (24°C)

Planting depth: 1/2–1 in. (1–2.5 cm)

Days to germination: 14–21 days

Spacing: 2–4 in. (5–10 cm)

Days to maturity: 55 days

Beet red does not apply here. 'Golden' beets are actually more of an apricot color when fresh, changing to a jewel-toned amber as they cook. Unlike heirlooms with romantic tales that are embellished over time, 'Golden' beets have a far more prosaic claim to fame: they do not "bleed" like red beets. When these beets were introduced, their color was a novelty, but the sweet flavor kept gardeners asking for more. These beets do not grow very large, but even when mature they retain their sweetness. The tops are tasty, too.

SWEET

Beet 'Golden'
Golden beet
Beta vulgaris

Flavor Succulent 'Golden' beets taste almost like fruit. One of the sweetest beets you will encounter, they have a smooth, fine texture that makes them a favorite for eating fresh or for cooking. Because they do not stain what they touch, as do red beets, they play well with other vegetables, both raw or simply roasted.

Growing notes Direct sow seed in the garden after the soil is dry and has warmed to 50°F (10°C). To maintain a steady harvest, plant seeds in succession every 3 weeks until midsummer. In the late summer or early fall, after nighttime temperatures have cooled off to at least 75°F (24°C), you can resume planting seeds for a fall harvest. Plant at least 1 month before the first expected frost date.

Soaking beet seeds overnight will soften their tough coating and speed germination. Tiny beet seeds come in clusters and they are difficult if not impossible to separate. Germination rates tend to be low for 'Golden' beet seeds, so oversow your rows; it means more thinning, but this will help you avoid disappointment, and you can add the thinnings to salads and soups.

Beet roots grow half in the ground and half above ground, so do not plant seeds too deeply and thin the plants to about 2–3 in. (5–7.5 cm) apart. They attain their sweetest flavor in slightly alkaline soils.

How to harvest 'Golden' beets are usually harvested when small and tender. Consider the size you prefer as you thin the plants; smaller sized beets need less room to grow than larger beets. You can thin out the fast growers and leave the slow pokes to develop.

Start harvesting the tops after they reach 2–6 in. (5–15 cm) tall, when they are most tender. The golden stems make an attractive side dish. Harvest the roots any time after they have reached about 2–3 in. (5–7.5 cm) in diameter. They will pull out of the ground easily, especially when they are small.

Others to try 'Albina Vereduna' is a white beet that is especially sweet when small. 'Yellow Cylindrical' is very sweet when small but not very tasty if they get too big. ('Golden Detroit' might be the same beet by a different name.)

BROCCOLI, CABBAGE, CAULIFLOWER, AND KOHLRABI

The cold-hardy family of cruciferous vegetables includes tender mouthfuls of tightly
closed flower buds; dense heads and loose
frills of crunchy, succulent leaves; and plump,
opulent roots. They get their daunting family
name from their four-petaled flowers that resemble small crosses. You will also see them
described as cole crops or members of the
mustard family, but that does an injustice to
their diversity.

These vegetables are not shy and retiring.
Strong in flavor and bite, they hold their own
with everything from wine to heady Indian
spices. You can chop them up for salads, use
them for dipping or stuffing, or cook them
until tender. Just get them started early in
the garden and keep the recipes coming
throughout the summer.

BROCCOLI, CABBAGE, CAULIFLOWER, AND KOHLRABI

I suspect you would be purple, too, if you stayed outside all winter. 'Early Purple Sprouting' broccoli is so hardy you can let it overwinter and it rewards you with—yes, you guessed it—early purple sprouts that are tender and delicious. The chilly temperatures give them an extra boost of sweetness.

As with so many purple vegetables, after cooking, only the young leaves remain colorful. Mix some of the tender leaves into your dish to help camouflage the broccoli for the kids.

Broccoli 'Early Purple Sprouting'

Brassica rapa (Italica Group)

Exposure:
Full sun to partial shade

Ideal soil temperature:
65–75°F (18–24°C)

Planting depth:
1/4–1/2 in. (0.6–1 cm)

Spacing:
8–10 in. (20–25 cm)

Days to germination:
7–14 days

Days to maturity:
60 days

Flavor After a winter of preserved or store-bought vegetables, fresh and uncooked 'Early Purple Sprouting' broccoli is a treat as a snack or in salads. As a cooked vegetable, the shorter the cooking time, the sweeter the sprouts; try stir-frying or blanching them.

Growing notes In cool climates, start broccoli indoors in the spring and transplant out shortly before the last forecasted frost. Young broccoli plants can be sensitive to cold weather. If you are direct seeding in areas with frost, sow seeds about 1 month after your last frost date. Gardeners in frost-free climates can direct seed in early spring. Gardeners in all areas can direct seed in midsummer for a fall harvest and to overwinter in the garden.

This is a wide plant, so give it considerable room. The plants start growing upward before stretching out their sizable leaves and forming a head. Spring-grown plants that have been kept in pots will grow more slowly and will be ready to transplant into the garden in late summer, when the main season plants are exhausted.

Harvest in fall and overwinter for later harvest.

Provide steady water, especially while the heads are developing. Side dressing midseason with compost or manure will give plants the boost they need to keep producing throughout the season.

Watch for cabbage family pests, such as cabbage looper, cabbage worm, and cabbage root maggot, which are at their worst early in the season. Add a row cover to keep them from chewing and destroying your plants. Fall plantings will not be bothered by pests.

How to harvest The sprouts grow in the axels between the stems. The more you harvest, the more the plants will sprout. If you wait too long to harvest the sprouts, they will eventually grow into full size heads, turning more green and growing less sweet as time passes.

Others to try 'Green Goliath' is an overachiever, with side shoots that appear for weeks after the initial head is harvested. 'Waltham' is a head-forming variety that rebounds with plenty of sprouts. 'White Sprouting' shoots resemble tiny cauliflowers; although it is not winter hardy, it is slower to bolt than the purple variety.

Cut and enjoy the first center head of 'Early Purple Sprouting' broccoli to encourage side shoots to form.

Looking at 'Romanesco' is as much a pleasure as eating this mesmerizing vegetable. The head is a cluster of bright green, spiraling conical florets, or peaks, which in turn spiral around the whole head in a fractal pattern. It looks somewhat alien and not at all like traditional broccoli. In fact, it is not a broccoli at all and is instead something of a broccoli-cauliflower. You can break apart the florets as you would with cauliflower.

'Romanesco' dates back to at least the 16th century, where it was first identified in Italy. It is still considered an Italian vegetable, although it is becoming more widely available at local farm stands.

Broccoli 'Romanesco'
Roman cauliflower
Brassica oleracea (Botrytis Group)

Flavor "Nutty" is the adjective most often used to describe the flavor of 'Romanesco'. When eaten raw, it is a bit spicier than broccoli, but it tastes much sweeter when cooked. The texture is interesting, too, more crumbly than crisp. Cook it fast and lightly; too much cooking will ruin the texture. Top with butter or a mild cheese sauce, or try traditional Italian seasonings.

Growing notes 'Romanesco' seed can be difficult to find, because it is sometimes listed under broccoli, sometimes cauliflower, and sometimes all by itself. Keep hunting. Seed can be started indoors, 5-6 weeks before the last expected frost date. Wait until all danger of frost has passed before transplanting seedlings outside. Gardeners in frost-free climates can direct sow seed in early spring. You can start seed in midsummer to harvest in fall and overwinter for later harvest.

Broccoli is not the easiest vegetable to grow, and 'Romanesco' can be especially fussy. It prefers partial shade and a slightly alkaline soil. This tall plant sometimes grows 3 ft. (0.9 m) before forming a head. It can also grow quite wide, so give it plenty of space.

The plants require steady water, especially while the heads are developing. Side dressing midseason with compost or manure will give them the boost they need to keep producing throughout the season. It will not resprout after harvest like a broccoli.

Watch for cabbage family pests, such as cabbage looper, imported cabbage worm, and cabbage root maggot, which are at their worst early in the season. A row cover will keep them from chewing and destroying your plants.

How to harvest Be patient while 'Romanesco' slowly matures. Wait until the head has fully formed and filled out before you harvest. You can harvest the whole head at once or break off small pieces.

Others to try There is nothing like 'Romanesco', but you can try 'Green Macerata', a large, apple-green cauliflower that is sweet and tender and forms a head earlier than 'Romanesco'.

Exposure:
Full sun to partial shade

Ideal soil temperature:
50–75°F (10–24°C)

Planting depth:
1/4–1/2 in. (0.6–1 cm)

Spacing:
2–3 in. (5–7.5 cm)

Days to germination:
7–14 days

Days to maturity:
160 days

As with so many oddly shaped vegetables, no two 'Romanesco' heads will look alike, and few will turn out picture perfect.

A broccoli without a head might sound silly, but try 'Spring Raab' once and you will appreciate it as one of the treasures of the spring garden. Like a cutting lettuce, this super fast grower keeps producing all season long.

Broccoli raab is more closely related to turnips than broccoli—in fact, rapa means turnip in Italian. The American version came about when an Italian immigrant stumbled on some wild plants in California that reminded him of the *rapa di broccoletti* from home. He began breeding and marketing the plants in the mid-1900s.

Broccoli 'Spring Raab'
Broccoli raab, 'Spring Rapini'
Brassica rapa (Ruvo Group)

Flavor American broccoli raab is more tender and less bitter than the Italian version. I am partial to 'Spring Raab' because a few flower buds are always interspersed with the leaves, giving it a blend of sweet and sharp flavors and an interesting texture. It tastes almost like an herb. The slight bitterness can add a zesty kick when mixed with milder or sweeter greens such as spinach and Swiss chard; it positively transforms tamer foods, such as potatoes. Of course, it also plays nicely with traditional Italian seasonings, such as garlic, olive oil, and red pepper.

Growing notes Start seeds indoors 6–8 weeks before the last expected frost date or direct seed outdoors after danger of frost has passed. Direct seed in midsummer to late summer for a fall harvest. Gardeners in frost-free climates can sow seed throughout the fall.

'Spring Raab' is a fast grower. You can get your seed or plants in the garden even while a chill is still in the air. If the weather is dry, keep them watered. Then keep taking cuttings so the leaves will keep growing. As a leaf crop, it appreciates some additional nitrogen, such as fish emulsion or soybean meal, while growing.

Although it matures quickly, it is susceptible to cabbage worms, flea beetles, and slugs. Because you do not need to worry about bees and pollination, floating row covers will protect the leaves from pest attacks.

How to harvest When it is time to harvest, raab should smell herbal and leafy. Past its prime, it will develop a cabbage-like smell. Cut outer leaves after the plants reach 4–8 in. (10–20 cm) tall. Harvest the flowers just as the buds begin to open. The stems can become tough and are often discarded. Harvest every few days or your plants will bolt to seed.

Others to try Often the only seed you will find is labeled simply as broccoli raab. 'Novantina' is a long-season Italian variety with large leaves.

Exposure:
Full sun to partial shade

Ideal soil temperature:
60–65°F (16–18°C)

Planting depth:
1/4–1/2 in. (0.6–1 cm)

Spacing:
1–3 in. (2.5–7.5 cm)

Days to germination:
6–10 days

Days to maturity:
40 days

As its name implies, 'Spring Raab' is most flavorful in the cooler months of the year.

The dense little bundles of 'Ching Chang' bok choy look as icy cool as the spring temperatures they love. This dwarf variety matures quickly and can be enjoyed before its tender, sweet bite is made bitter by heat.

Bok choy (Chinese for white vegetable) is a loose heading cabbage. Although it has been cultivated for centuries in China, it made its way to Europe in the 1800s. 'Ching Chang' stalks are a bit greener than those of the average bok choy, which seems to make it more heat tolerant and slower to bolt.

Bok choy 'Ching Chang'

Chinese cabbage, pak choy
Brassica rapa (Chinensis Group)

Flavor The crinkled green leaves have a sweet tenderness similar to that of young Swiss chard greens. The leaves are complemented by the slightly bitter crunch of the firm and succulent white stalks. The stalks are short-er than those of a standard bok choy, so they do not overwhelm the greens. 'Ching Chang' can be a quick substitute for any type of cabbage and should not be relegated to Asian dishes only. Try using them to make cabbage rolls or with corned beef.

Growing notes Direct seed in the garden after the last spring frost date. You can also start seedlings indoors about 4–6

Exposure:
Full sun to partial shade

Ideal soil temperature:
50–75°F (10–24°C)

Planting depth:
1/4–1/2 in. (0.6–1 cm)

Days to germination:
4–7 days

Spacing:
6–8 in. (15–20 cm)

Days to maturity:
55–60 days

Given the right conditions, fertile soil, cool weather, and plenty of water, 'Ching Chang' will quickly grow into stout little heads.

weeks before transplanting out. In frost-prone climates, you can sow a second crop in late summer to mature in the fall. Gardeners in frost-free climates can grow 'Ching Chang' fall through winter.

This heat-resistant bok choy can be sensitive to cold temperatures below 50°F (10°C), so do not start plants too early. Because it is grown for its leaves, a soil rich in organic matter will make for more lush plants.

All bok choy varieties require regular watering to remain succulent. Mulching plants will keep the roots cool and moist. Make sure the soil is well draining or the roots can rot.

You can side dress with compost or manure 3–4 weeks after sowing seeds, but since 'Ching Chang' grows so quickly, soil amendment is not mandatory.

Watch for snails and cabbage worm eggs. Hand remove the eggs and use *Bacillus thuringiensis* (Bt) insecticide when necessary.

How to harvest Start harvesting when the leaves reach 5 in. (12.5 cm) tall. Either pull off a few outer leaves or cut off the whole plant at the soil line.

Others to try Baby bok choy is similar in size and flavor but quicker to bolt. 'Extra Dwarf' pak choy are tiny, 2–3 in. (5–7.5 cm) plants that can be cooked and served whole. 'Purple Stalk' is larger, with lavender-hued stalks and more cabbage flavor.

Exposure:
Full sun to partial shade

Ideal soil temperature:
65–75°F (18–24°C)

Seed Planting depth:
1/4–1/2 in. (0.6–1 cm)

Days to germination:
5–20 days

Spacing:
2–3 ft. (0.6–0.9 m)

Days to maturity:
90 days

You can recognize savoy cabbages by their crinkled leaves. The leafy parts grow faster than the veins, eventually folding over on themselves, resulting in crunchier, more tender leaves than a smooth-leaved cabbage. Crisp and firm, yet tender and refreshing, 'Perfection Drumhead Savoy' holds its leaves tightly, protecting its cool crunch.

This cabbage is ideal for gardens of any size. Unlike sprawling cabbage heads whose inedible outer leaves take up space, this cabbage forms a compact head on a sturdy, short stem. It was introduced in Paris in the winter of 1888, where it reportedly was shipped into town by the truckload. It gets its name from its somewhat flattened top.

Cabbage 'Perfection Drumhead Savoy'
Savoy cabbage
Brassica oleracea (Capitata Group)

Flavor A good head of 'Perfection Drumhead Savoy' will be firm and heavy, a delight to chop. It has a sweet, mild cabbage flavor; when gently cooked, it releases less of that musky scent associated with the typical cabbage. For cole slaw and other salads, it offers crisp cabbage flavor without overwhelming the dish.

Growing notes Wait until temperatures stay above 50°F

(10°C) before placing cabbage plants outdoors. You can start seeds indoors 4–5 weeks earlier. Seedlings that have been hardened off can handle cool spring temperatures. Getting your transplants into the garden in early spring ensures they will mature before the summer heat. To avoid having dozens of cabbages ready to use all at once, start seeds a few weeks apart in the spring to stagger the harvest.

You can start new cabbage seedlings in late summer, but the heads will not begin to form until the weather cools. In frost-free climates, cabbage can be grown throughout the winter. They can be direct seeded or started indoors.

'Perfection Drumhead Savoy' is a heavy feeder and needs rich soil, but do not overfertilize or the cabbages will crack. Cabbage does not do well in acidic soils, so add some lime if your soil pH dips below 6.5.

Because the cabbage does not need to be pollinated, row covers can be used to protect the leaves from cabbage pests such as loopers, cabbage worms, and slugs.

How to harvest Harvest as soon as the head feels firm. Cut the stem below the head. Do not wait too long to harvest or the heads will split; if heavy rain is expected, harvest before the rain. Discard the floppy outer leaves.

Others to try 'January King' is a purple-tinged semi-savoy that can handle the cold. 'Red Verone Savoy' has maroon-blushed leaves and a mild flavor. 'Savoy Chieftain' has blue-green leaves and a great puckered texture.

'Perfection Drumhead Savoy' is so dense and wrinkled that water drops form puddles on its leaves.

Until you see something as beautiful as 'Di Sicilia Violetto', you might forget that a head of cauliflower is really a flower. This Sicilian cauliflower looks as though it was washed in a soft violet dye. It is lovely when fresh, but the curds turn green, as do so many purple vegetables, when cooked. Thankfully, it retains all of its sweetness after cooking.

Italian vegetables tend to be regional, and the names are a succinct summation of the variety and town. 'Di Sicilia Violetto', or Violet of Sicily, is well suited to Sicily's sunny, mild climate. American gardeners who want to indulge in this Italian delicacy will fare better growing it in the fall rather than planting it in the spring.

BEAUTIFUL

Exposure:
Full sun to partial shade

Ideal soil temperature:
65-75°F (18-24°C)

Seed Planting depth:
1/4-1/2 in. (0.6-1 cm)

Days to germination:
5-20 days

Spacing:
18-24 in. (45.5-61 cm)

Days to maturity:
80-90 days

Cauliflower 'Di Sicilia Viletto'
Brassica oleracea (Botrytis Group)

Flavor The flavor of this cauliflower is similar to that of broccoli, but milder and sweeter, with more nutty richness. The trick with 'Di Sicilia Violetto' is not to overcook it. If you zap it in the microwave, without water, until it is almost done, it may retain its purple color. Then let it sit uncovered to finish cooking through. No need to tell the kids that this festive-looking vegetable is rich in vitamins and nutrients.

Growing notes Transplant seedlings outdoors after all danger of frost has passed. Seedlings can be started indoors about 4-6 weeks earlier. 'Di Sicilia Violetto' matures quickly in the spring. Gardeners in cooler climates may have better luck sowing in midsummer to late summer, to harvest in the fall. Gardeners in frost-free climates will have the best success growing this cauliflower in the fall through winter.

Set plants outside only when conditions are best for growing. Wait until temperatures remain reliably above 55-65°F (13-18°C) and the soil warms to 65-75°F (18-24°C). If plants are set in cold soil, they will start forming heads prematurely and can be smaller than expected.

Cauliflower is not an easy vegetable to grow well, and home gardeners generally get small heads, but its tender sweetness more than makes up for its size. They are greedy feeders, so start with a rich soil and either work in a slow-release organic fertilizer or feed the plants every 2 weeks with a high nitrogen fertilizer. The violet heads mean that no blanching or tying up the heads is required.

All kinds of pests are waiting to devour your cauliflowers. Cabbage worms and cabbage root maggots are among the most destructive. Row covers will keep them out. Club root swells the roots and underground stem and prevents the plant from taking up nutrients. The organisms that cause it remain in the soil, so rotating your brassicas is the best way to avoid this.

How to harvest Cut the whole head when it feels firm but the surface is still smooth. Do not wait until it develops a grainy texture.

Others to try 'Early Snowball' is a tasty, compact, short-season cauliflower. 'Garantiya' (which means guarantee in Russian) is suitable for shorter growing seasons. 'Lilovyi Shar' is a purple Russian variety with good cold tolerance.

The color of 'Di Sicilia Violetto' can vary from a soft violet to a purple haze.

One bite will tell you that 'Early Purple Vienna' kohlrabi is in the cabbage family. It looks beautiful, cool, and mouthwatering—and it is. This vegetable has been grown for a century and a half, but most of us have never met it or have shied away from it. Set out a few slices with a dip and watch people reach for more.

We eat the bulb part of the plant—or, more precisely, the swollen stem. The leaves are also tasty, but the crunchy bulb is the prize. Kohl is Latin for stem and also means cabbage in German, and rabi is a derivative of rapa, which refers to turnips. Kohlrabi, however, is not a cross, but a respectable vegetable on its own.

Exposure: Full sun to partial shade

Ideal soil temperature: 65–85°F (18–30°C)

Planting depth: 1/4–1/2 in. (0.6–1 cm)

Days to germination: 5–15 days

Spacing: 6–8 in. (15–20 cm)

Days to maturity: 55–65 days

Kohlrabi 'Early Purple Vienna'

German turnip

Brassica oleracea (Gongylodes Group)

Flavor This kohlrabi is a succulent stem with a flavor that is both complex and refreshing, with hints of broccoli and cucumber and a juicy sweetness. The flavor of this selection is more robust than paler varieties of kohlrabi, and its pleasant purple appearance is far less intimidating for the uninitiated.

Move over celery, because kohlrabi should be on every crudité plate. Cooking brings out its sweetness, and it is especially nice just heated through, al dente. It can also be chopped and used in stuffing and added to soups and stews.

Growing notes Plant transplants as soon as the soil can be worked, or start seed indoors 6–8 weeks earlier. Plant a fall crop in late summer. Summer-sown kohlrabi will need some shade to get started, but it will be even sweeter when harvested in the fall.

Kohlrabi should grow at a fairly steady rate, so start it off with a rich soil and lots of moisture. Once it is established, water only when the soil gets dry to keep the stem from splitting, but do not let it remain dry for long periods.

In rich soil, supplemental fertilization should not be required, but you can amend the soil with more compost before starting another crop.

'Early Purple Vienna' is relatively pest-free, but all kohlrabi varieties are subject to the usual cabbage family pests, such as cabbage loopers and cabbage worms. If these pests are a nuisance in your area, you can grow it under a row cover since it needs no pollination.

How to harvest Harvest young, when the stem bulb is only 2–3 in. (5–7.5 cm) in diameter, by cutting just below the bulb. If kohlrabi grows too large, it will become woody and split, so plant a lot and enjoy them while they are still tender.

Tip Unlike most store-bought kohlrabi, home-grown bulbs are fresh, tender, and need no peeling, which makes eating 'Early Purple Vienna' a real treat.

Others to try 'Gigante' is a Czechoslovakian heirloom and

world record holder for the largest kohlrabi; it is mild and stays tender. 'Grand Duke' is more uniformly round. 'Violeta' is a late-maturing variety with good cold tolerance.

Kohlrabi 'Violeta' hovers over the ground like a little alien space ship.

CARROTS, CELERIAC, AND FENNEL

All members of the family Apiaceae (formerly Umbelliferae) feature lacy, umbrella-shaped flowers like their wild cousin, Queen Anne's lace (Daucus carota). This family includes many herbs such as caraway, cumin, and dill as well as some deceptively sweet roots and bulbs. Some, such as carrots and celery, are common sights that can show up in unexpected colors and shapes. Others puzzle us but are destined to become favorites, after we figure out how to flaunt their delectable idiosyncrasies. You need to be patient as they work their wonders, hidden away underground, but they are well worth the wait.

'Dragon' carrot is decidedly sweet, but it sneaks up on you with its warm, spicy finish. Reddish-purple with an orange-yellow center, this purple dragon has a hint more zest than the average orange carrot.

'Dragon' might look like a designer vegetable, but the first cultivated carrots were purple and they remained popular for centuries. It requires more time and care in the garden than mass-marketed carrot varieties, so it is a pleasure reserved for the home garden. This flushed carrot is said to contain levels of the antioxidant lycopene equivalent to those found in tomatoes.

Carrot 'Dragon'
'Purple Dragon'
Daucus carota

Flavor 'Dragon' is sweet and crunchy, and its striking colors make it popular as a fresh carrot. It tastes almost like cinnamon and roasts wonderfully, enhancing its spicy nature.

Growing notes These long carrots need a loose, tilled soil to grow straight and deep. Rocks or compacted soil will cause carrots to deform. They also need constant moisture. Plenty of organic matter in the soil will help with both.

Direct sow seeds in the garden. In Zones 7 and colder, sow carrots in early spring, 3–4 weeks before the last expected frost, and succession plant until midsummer. In frost-free climates, carrot seeds can be sown any time from fall through early spring.

The seeds are tiny and difficult to work with, so do not worry about spacing; you can thin the seedlings later. Water them well with a fine spray, and then mulch lightly to keep the soil cool and moist.

Start thinning the seedlings when they are about 1–2 in. (2.5–5 cm) tall. To avoid disturbing the remaining carrots, remove thinnings by snipping the tops at the soil line. Thin again when you start to see the shoulders of the carrots forming just below the soil. You can probably harvest some baby carrots that are large enough to eat from the second thinning.

How to harvest Harvest by twisting and pulling the stems, or use a fork to loosen the soil and then pull the carrot out of the ground. It can be difficult to judge when carrots are ready to harvest. 'Dragon' is a long-season carrot, at 90 days, but you can start testing your crop a week or two earlier. Pull up one or two and taste. If they taste sweet and spicy, they are ready to harvest. Do not let them stay in the ground too long once they are full grown or they get woody.

A late planting might not mature before the first frost hits, but they can be left in the ground and harvested the following spring.

Tip Remove the tops of carrots before storing them; the tops continue growing, sapping moisture and flavor from the carrot roots.

Others to try 'Atomic Red'

Exposure:
Full sun

Ideal soil temperature:
65–75°F (18–24°C)

Planting depth:
1/4–1/2 in. (0.6–1 cm)

Spacing:
2–3 in. (5–7.5 cm)

Days to germination:
12–15 days

Days to maturity:
90 days

Beet red outside and golden orange inside, these carrots offer a surprise with a spicy bite.

carrots are high in lycopene and blazing red, with a sweet and crisp flavor. 'Cosmic Purple' is another spicy carrot with purple skin and yellow-orange interior. 'Danvers' is a sweet, dark orange variety that is a good choice for gardens with heavy soil.

Their name may conjure up thoughts of lean and sophisticated haute cuisine, but at 1–2 in. (2.5–5 cm) long, 'Paris Market' carrots are cute little dumpling carrots and a tender, sugary mouthful. They grow quickly for a carrot, which probably explains their sweetness. Their tiny size and quick growth also make them perfect for shallow soil, containers, and repeat harvesting.

'Paris Market' also goes by the names 'Tonde de Paris' and 'Parisien Market'. This carrot is so popular that hybrid baby carrots are often compared to 'Paris Market' in seed catalogs.

Snack-sized 'Paris Market' is perfect for growing in containers and is a quick grower.

SWEET

Carrot 'Paris Market'
'Parisien Market', 'Tonde de Paris'
Daucus carota

Flavor These are not baby carrots, harvested before their sugars develop, or the bland, shaved-down carrots often sold as baby carrots. The concentrated sugar content of 'Paris Market' reaches the peak of its crisp sweetness while the carrot is the size of a nugget. These tiny carrots cook quickly and retain all their candy sweetness. Although they make a tempting fresh snack, cooking them until just tender helps caramelize the sugars.

Growing notes In Zones 7 and colder, direct sow seeds in early spring, 3–4 weeks before the last expected frost, and succession plant throughout summer. In frost-free climates, carrots can be seeded from fall through early spring. The seeds are tiny and difficult to work with, so do not be too concerned about spacing. You can thin the plants later.

Water them well with a fine spray, and then mulch lightly with straw to keep the soil cool and moist. Start thinning the seedlings when they are about 1–2 in. (2.5–5 cm) tall. To avoid disturbing the remaining carrots, thin by snipping the tops at the soil line. 'Paris Market' does not need much room to spread out or down, making it perfect for gardens with hard or rocky soils.

How to harvest You can start harvesting shortly after you see the carrots' shoulders pushing up soil. Harvest by twisting the stems and pulling them out of the ground.

Tip Try interplanting 'Paris Market' with flowers and in containers. The wispy foliage provides a nice contrast against bold-leaved plants. By the time the larger plants fill in, the carrots are ready to harvest.

Others to try 'Early Scarlet Horn' is a full-sized carrot with red shoulders that is wonderfully sweet when harvested young. It is thought to be the oldest cultivated carrot. 'Little Finger' is another quick, sweet baby carrot from France that is great for pickles. 'Red Core Chantenay' is a somewhat stumpy carrot that is extremely sweet and makes great juice.

Exposure:
Full sun

Ideal soil temperature:
65–75°F (18–24°C)

Planting depth:
1/4–1/2 in. (0.6–1 cm)

Spacing:
2–3 in. (5–7.5 cm)

Days to germination:
12–15 days

Days to maturity:
50–65 days

Crunchy, with an apple-sweet juiciness, 'Giant Prague' celeriac is a succulent, swollen stem and a celery cousin. It has to be tasted to be appreciated, because no one would be attracted by its looks alone. The leafy stalks look familiar enough, but the part we eat—the juicy, fleshy bulb—grows below ground and is gnarled with misshapen roots running in all directions. 'Giant Prague' is the most uniform and reliable of heirloom celeriacs and the most widely grown. It also has all the concentrated flavor of a fruity celery.

Celeriac traveled from Egypt to Britain in the 18th century, where it went unappreciated. Central Europeans, however, always on the lookout for a winter vegetable, latched onto it and made it their own, adding it to everything from salads to desserts. 'Giant Prague' was introduced in North America in the early 19th century and is finally appearing on the menus of American cooks.

Celeriac 'Giant Prague'

Knob celery

Apium graveolens var. *rapaceum*

Flavor Celeriac's flavor falls somewhere between celery and parsley, with an earthy warmth that reminds you it is an underground bulb. Freshly cut, it is chewy and intense. Marinating softens the flavors and tenderizes the texture. But 'Giant Prague' is at its best when it is used in stews and soups, where it is allowed to lend its flavor and combine with others.

Growing notes 'Giant Prague' requires a long growing season. Start seed indoors, 8–10 weeks before transplanting outside after all danger of frost has passed. In frost-free areas, it can be grown or seeded in the fall. Soak seed overnight to aid germination. Press the seed on the surface of the soil, but do not cover. Keep the soil moist until seeds germinate.

'Giant Prague' will spend several months in the ground, so a good start is crucial. Provide a soil amended with lots of organic matter. To help the plants thrive, side dress with compost or feed with a balanced fertilizer monthly and give them regular water.

Slugs can attack young plants, but carrot and celery flies can be the most troublesome pests. Row covers will prevent the flies from landing on the plants or soil and laying their eggs. Lessen the problem by planting your celeriac in a different spot each year, in case the larvae overwinter in the ground near the current crop.

How to harvest You can space seedlings close when transplanting, and then thin and use alternate bulbs when they reach at least 2 in. (5 cm) in diameter. Leaving the bulbs in the ground until the tops are hit by frost will make them sweeter.

Tip Because celeriac is a biennial, you need to harden off the plants slowly. If they are exposed to cold temperatures too

Exposure:
Full sun

Ideal soil temperature:
70–75°F (21–24°C)

Seed planting depth:
Press seed on soil surface

Days to germination:
14–21 days

Spacing:
6–8 in. (15–20 cm)

Days to maturity:
110–120 days

Knobby 'Giant Prague' celeriac might look odd, but it offers a dense package of sweet celery flavor.

suddenly, they will start going to seed when the temperature warms up.

Others to try 'Apple-shaped' is small, smooth, and early, but almost impossible to find. 'Brilliant' is a mellow, early, and smooth-skinned variety. 'Early Erfurt' is a flavorful, medium-size bulb with a uniform shape.

Savory and sweet, 'Zefa Fino' fennel delivers aromatic flavor from all parts of the plant: stalks, foliage, and seeds. Fennel seed is one of the most widely used spices in the world and is used to season everything from cough drops, to sausage, to chai tea. 'Zefa Fino' consolidates all that flavor into a mouth-watering vegetable.

Fennel, or finocchio, is usually associated with Italy, where it has been cultivated since the 17th century. Thomas Jefferson was probably the first American to be sent some seeds to try. He loved it, but fennel never really caught on in America. 'Zefa Fino' was bred in Switzerland more than 100 years ago. It is a compact, vase-shaped plant, with less airy foliage than many modern fennels, which means this old variety puts most of its energy into growing a plump, tasty bulb.

AROMATIC

Fennel 'Zefa Fino'

Finocchio, bulb fennel, Florence fennel

Foeniculum vulgare var. *azoricum*

Flavor Fennel is often compared to anise, which it is not, and licorice. 'Zefa Fino' has a more complex and mellow flavor than typical fennel, with just a touch of warmth to balance the sweetness. When eaten raw, it offers the crunch and juiciness of celery hearts. It really delights when it is roasted and the aroma combines with the caramelized sugars to create an ambrosial feast.

Growing notes 'Zefa Fino' is a cool season crop and can be direct seeded in early spring, but seedlings will need protection from late frosts. Gardeners in Zones 7 and cooler can start seed indoors 4–6 weeks before the last expected frost, to plant out after danger of frost. If you are starting seed indoors, use a peat or paper pot, because fennel develops a tap root that does not transplant well. Transplant as soon as possible to avoid disturbing the tap root. Whenever possible, direct seed in the garden. You can succession seed into midsummer.

In frost-free climates, seed can be direct sown anytime, but 'Zefa Fino' will be most bolt-resistant if planted in the fall. Cooler climates can start a second planting in midsummer to mature in the fall.

'Zefa Fino' is a hungry plant. Start with a rich soil and feed it every 3–4 weeks with an all-purpose fertilizer or a side dressing of compost. Water is a crucial ingredient, and the plants may need watering a few times per week during hot spells. Mulching will help to conserve moisture.

How to harvest Start harvesting when the bulbs reach the size of a tennis ball. 'Zefa Fino' is hardy enough to survive some frost. If the plants are left to go to seed, they will self-sow throughout the garden.

Tip 'Zefa Fino' makes a beautiful container plant, but be sure

Exposure:
Full sun

Ideal soil temperature:
60–65°F (16–18°C)

Seed Planting depth:
1/4–1/2 in. (0.6–1 cm)

Days to germination:
10–14 days

Spacing:
12–24 in (30.5–61 cm)

Days to maturity:
80–85 days

'Zefa Fino' bulbs tend to be a bit flatter than most of the fennel bulbs you find in the produce aisle.

to use a pot large enough to provide a consistently moist soil. Swallowtail butterflies love it.

Others to try Most varieties of bulbing fennel are named some variation of Florence fennel, and they all are similar in taste and looks. You will also see 'Fino' and 'Finocchio Romanesco'.

COOKING GREENS

Most cooking greens are close relatives of cabbage, but these plants do not form dense heads. Instead, cooking greens are grown for their profuse production of leaves, which can be cooked into a diversity of dishes, from the most humble and homespun meal, to an elegantly upscale side dish.

Greens have been grown for thousands of years and all over the world. Many originally grew in the wild and are easy to grow in the garden. Collards, for example, a staple of Southern cooking in the United States, probably originated in the Mediterranean region, as did kale and Swiss chard (or silverbeet). They have also become popular in cooler climates because of their frost tolerance. Take a tip from the ancient Romans and enjoy your first harvest of greens by eating the seedlings you thin out at the start of the season. These tiny leaves offer a mellow savoriness and delicate texture.

Imagine seeing a marbled mosaic of amaranth leaves on your plate and finding it delicious, too. Used as a seed-like grain, a green, and an ornamental plant, amaranth is one of the most widely grown plants in the world, yet many gardeners are unfamiliar with how to use it. High in nutrients and flavor, these flamboyant plants with splashes of green, red, and yellow are slowly making a big splash in modern cuisine.

'Red Leaf' amaranth has been hiding in our flower gardens for years. Each dark green leaf is splashed with red down the center, similar to a coleus leaf. You may do a double-take when you see it on your plate, but 'Red Leaf' amaranth lives up to its reputation as a summer spinach. This prolific, heat-loving plant is a vitamin-rich green that can be grown and enjoyed all summer long.

Amaranth greens are thought to have originated in Central or South America, and they were a part of Native American diets. 'Red Leaf' is often associated with Chinese cuisine, but you can enjoy the greens mixed with spinach and beet greens and use them interchangeably in recipes. Although closely related to spinach and beets, amaranth is generally classified in its own family, Amaranthaceae.

LONG SEASON

Exposure:
Full sun

Ideal soil temperature:
70–75°F
(21–24°C)

Planting depth:
1/8–1/4 in.
(0.3–0.6 cm)

Spacing:
6–12 in. (15–30.5 cm)

Days to germination:
5–15 days

Days to maturity:
65–80 days

Amaranth 'Red Leaf'
Chinese spinach, Joseph's coat
Amaranthus tricolor

Flavor 'Red Leaf' keeps its fresh taste and texture when stored for up to two weeks. It is deliciously tender and chewy with a grassy, earthy flavor, similar to a full-bodied spinach. Young leaves can be used fresh in salads. Steaming allows older leaves to retain their texture and color, while releasing their flavors. Sauté with garlic and olive oil for a colorful side dish or pasta topping.

Growing notes Direct seed in the garden, 2–3 weeks after the last frost date or any time until midsummer. It does not transplant well. 'Red Leaf' amaranth is hardy to Zone 8 and can be succession planted every 2–3 weeks to provide a continuous supply of young leaves. Although it is a heat lover and grows faster in warm temperatures, it will germinate in cooler spring soil. It is a heavy feeder and likes a nitrogen-rich, organic soil.

The seeds are tiny and difficult to work with, so do not worry much about spacing. Do your best to sow the seeds about 2–3 in. (5–7.5 cm) apart and thin the seedlings when the plants are 1–2 in. (2.5–5 cm) tall. You can use the thinnings in soups and salads. Provide plenty of water, especially while the plants are young. Plants can tolerate drought but are more robust if kept watered.

Left unharvested, 'Red Leaf'

can grow taller than 5 ft. (1.5 m), but with frequent cuttings, the plants remain low and bushy. The seeds are also edible and are a source of protein.

How to harvest Harvest the leaves as needed. Snip the stem, but allow at least two sets of leaves to remain, so the plant can regrow.

Tip To differentiate leafy green varieties of amaranth, it is often listed under ethnic names, such as callaloo (Caribbean), quintonil (Mexican), and yin choy (Chinese).

Others to try Amaranth names can be colloquial and confusing, but some excellent, widely available varieties include 'All Red', with beautiful, tender, mild red leaves; 'Kerala Red', tangy enough to add flavor to curries; and 'Merah', with brilliant, yellow-trimmed burgundy leaves and a nutty flavor.

You might have seen amaranth in ornamental gardens, but have you tried it as a side dish?

Exposure:
Full sun to partial shade

Ideal soil temperature:
70–75°F (21–24°C)

Seed Planting depth:
1/4–1/2 in. (0.6–1 cm)

Days to germination:
5–10 days

Spacing:
12–18 in. (30.5–45.5 cm)

Days to maturity:
65–75 days

'Vates' is a rambunctious grower that sends up large, thick leaves.

'Vates' collards have succulent blue-green leaves that wave and glisten in the breeze. Originally, "collards" referred to the thinned plants of kale or cabbage. By the late 18th century, distinct varieties of collards developed from accidental crosses of the two plants, and most varieties are named for the region in which they were grown.

'Vates' collards were developed at the Virginia Truck Experiment Station, a forerunner of the agriculture research center. The station developed several varieties, but 'Vates' is most ideally suited to home gardens, being compact and tolerant of many growing conditions, including neglect.

PROLIFIC

Collards 'Vates'
Collard greens
Brassica oleracea (Acephala Group)

Flavor Although you will detect a cabbage accent when tasting 'Vates' collards, they are milder and sweeter than cabbage. They have the crisp crunch of cabbage and a sweetness reminiscent of kale, which is enhanced after a frost. Unlike the spicy bitterness of mustards, 'Vates' has a rich earthiness that develops as it cooks. It takes a "mess o' greens" to make a meal, because they cook down to a fraction of their raw size.

Growing notes Seeds can be started indoors, about 6–8 weeks before transplanting into the garden. Seedlings can handle a light frost and can be transplanted 2–4 weeks before your last expected frost date. Cooler climates can get a second harvest in the fall if plants are started in midsummer. Gardeners in frost-free climates can direct seed 'Vates' in the garden, fall through winter.

Plant transplants deeply, covering part of the stem. A rich, fertile soil and plenty of water will produce the most succulent leaves.

Plants can get large and bushy, but you can thin and eat the closely spaced seedlings throughout the season to make room along the way for more growth. Because the flavor improves with a touch of frost and 'Vates' is a hardy green, it often overwinters even in snowy climates, especially with a little protection.

Other than slugs and snails, few pests or diseases bother collards, and row covers will keep the leaves pristine.

How to harvest Cut the leaves as needed or slice off the entire plant after it reaches 6–10 in. (15–25 cm) tall.

Tip Collards are steeped in folklore. Eating collards, black-eyed peas, and hog jowls on New Year's Day is said to bring good fortune all year, and hanging a collard leaf over your front door will keep away evil spirits.

Others to try 'Georgia Southern Creole' is an old Southern standard, also known as 'Georgia Green', 'Georgia Southern', or 'Creole', with sweet, succulent leaves. 'Morris Heading' has a thick leaf but is a more finicky grower. 'Southern Greasy' has glossy, smooth leaves that almost look greasy but taste great.

The Italian word *laciniato* means curly, and the puckered leaves of lacinato kale can look a lot like pin curls. It is this heavily savoyed crinkle that gives this kale its great texture and allows it to hold up to frost, which sweetens its flavor.

Lacinato kale dates back to the 18th century and it is still one of the most widely grown kales in Italy. It goes by many names, including 'Nero di Toscana', 'Tuscan Black Palm', 'Dinosaur', and simply Tuscan kale. Its imposing stature and pebbly looking leaves do not prepare you for its meaty, mellow flavor, even when eaten raw.

LONG SEASON

Kale, lacinato

Tuscan kale, 'Nero di Toscana', 'Tuscan Black Palm', 'Dinosaur'
Brassica oleracea (Acephala Group)

Flavor If you are familiar with bitter, curly kale, you will be delightfully surprised by lacinato's flavor. Lacinato has a sweet, earthy crunch that is best lightly cooked so it retains some bite. Blanching before braising will sweeten the flavor. This hearty green is great with savory fare, such as sausage or ham, or

Exposure:
Full sun

Ideal soil temperature:
70–75°F (21–24°C)

Planting depth:
1/4–1/2 in. (0.6–1 cm)

Spacing:
16–20 in.
(40.5–51 cm)

Days to germination:
5–8 days

Days to maturity:
55–65 days

The long, crinkled leaves of lacinato kale give the plant a gangly appearance but a great texture.

blended with the sweet touch of currents.

Growing notes Direct seed 4–6 weeks before the last expected frost date, or start indoors at about the same time and transplant seedlings after danger of frost has passed. You can plant succession crops, because kale sprouts and grows quickly in warm weather. You can also direct seed in late summer to early fall. In cool climates, seeding lacinato in midsummer should allow it to mature just in time to be touched by frost in the fall. Gardeners in frost-free areas can succession plant into winter. Winter crops can taste much sweeter than summer crops.

Keep the seeds moist until they germinate. Lacinato's thick leaves are full of moisture, and the plants need to be kept well watered throughout the growing season or they will get tough and unpalatable. Mulching to keep the soil cool will also help retain its sweetness and tender crunch. Side dress with compost or manure or feed with a balanced fertilizer in midsummer.

Slugs do not usually bother lacinato's thick leaves, but cabbage worms will hatch and munch holes in them. Watch for the appearance of eggs and remove them by hand before they hatch, or use a row cover to prevent worms from eating the plants.

How to harvest Harvest small, young leaves to use fresh in salads. Harvest the outer leaves as needed and allow the plant to grow more. The leaves are their most tender when harvested at under 24 in. (61 cm) long.

Others to try No other kales are comparable to lacinato, but you will find some excellent curled or Scotch kales. 'Dwarf Blue Curled' is a tender baby kale best planted in late summer. 'Savoy' is an early spring variety, a cross between 'Dwarf Scotch' and lacinato, with a sweet flavor. 'Vates Blue Curled' is an old standard that is compact and overwinters well.

Kale is one of the prettiest vegetables to grow, and 'Red Russian' is a show-stopper. The frilly blue-green leaves with purple veins and stems make them a study in cool colors. 'Red Russian', as the name implies, comes from Russia, and it can handle cold weather. It not only gets sweeter, but its color intensifies in the cold.

This quick and easy grower stays tender, mild, and sweet throughout summer. It has had some less than flattering names in the past, such as 'Ragged Jack' and 'Communist Kale'. The little tags that grow out of the leaf surface make 'Ragged Jack' an understandable moniker. 'Red Russian' was said to have been introduced to Canada by Russian traders, before making its way down to the United States, but I do not think the plant ever had any political affiliation.

Kale 'Red Russian'

Brassica oleracea (Acephala Group)

Flavor One of the mildest flavored kales, it is more earthy than tart and acts as a sponge or bridge between other flavors. Being a flat-leaf kale, it is more tender than most and requires less preparation and cooking. Its chewy crunch and subtle spice are great in soups, sautés, or mixed in with other vegetables such as broccoli or beans.

Growing notes Direct seed or start indoors 4–6 weeks before the last expected frost date. Because kale sprouts and grows quickly in warm weather, gardeners in all climates can direct seed in late summer to early fall. Gardeners in frost-free areas can succession plant into winter. Winter crops can actually taste much sweeter than summer crops.

Start with a soil rich in organic matter, and keep the seeds moist until they germinate. Mulch seedlings to keep the surrounding soil cool to help retain its sweetness and tender crunch. Water is also crucial for keeping the leaves tender. 'Red Russian' will wilt in high heat and humid-

Exposure:
Full sun to partial shade

Ideal soil temperature:
70–75°F (21–24°C)

Planting depth:
1/4–1/2 in. (0.6–1 cm)

Spacing:
12–18 in. (30.5–45.5 cm)

Days to germination:
3–10 days

Days to maturity:
50–60 days

Any vegetable with the word *Russian* in its name will be cold-hardy; 'Red Russian' kale keeps producing long into winter.

ity, so some shade and plenty of water is important, especially during hot spells. Side dress with compost or manure, or feed with a balanced fertilizer in midsummer.

Cabbage worms and slugs can be a problem, and as with most brassicas, row covers will keep these pests at bay.

How to harvest Pick the small, young leaves to use fresh in salads. Harvest the outer leaves as needed and to allow the plant to grow more. A late-season crop makes wonderful baby kale.

Tip Plunge leaves into ice cold water before storing to keep them crisp.

Others to try 'Gulag Stars' is another flavorful Russian kale with leaves of different colors and textures. 'Russian Frills' is a frilly version of 'Red Russian' that can be difficult to find, but the texture is great. 'Siberian' is similar, but without red veining and stems. It is tender, slow to bolt, and great for overwintering.

Swiss chard is not really chard at all; it is in fact a type of beet. The name is derived from carde, the French word for cardoon. Something was lost in the translation, and Americans now use the name Swiss chard interchangeably with silverbeet. You may also find it listed as sea kale beet or spinach beet.

You will have no trouble getting your family to eat cooked greens if you serve this colorful Swiss chard. The flavor is both sweet and bitter, earthy and light. Leaf stems and veins are colored deep red, pink, gold, and orange, and leaves are various shades of green.

Maybe it is the beauty of this Swiss chard that once made it more popular in an ornamental garden than on the dinner table. To keep the colors so bright, plants of each color have to be grown separately by seed growers; then the seeds are mixed together and sold as a blend for planting. This plant is a biennial: plants take two years to produce seed. Although it is easy to grow, a lot of effort went into preparing the seeds for you.

COLORFUL

Swiss chard 'Rainbow Chard'

Five-color silverbeet, 'Bright-Lights'
Beta vulgaris subsp. *cicla* var. *flavescens*

Flavor Swiss chard leaves offer a tangy sweetness with a bitter spiciness when cooked. The flavor is more intense than spinach, but not as bitter as kale or collards. Use the young, leaves fresh in salads, or sauté the greens with garlic and a splash of lemon and serve topped with parmesan cheese.

Growing notes Direct sow in the garden, 2–3 weeks before the last expected frost. You can also start seed indoors about 5–6 weeks earlier and transplant the seedlings into the garden. Gardeners in frost-free areas can sow throughout the fall and winter and probably into summer.

Tiny Swiss chard seeds come in clusters, similar to beet seeds. You can either gently crush the seeds apart before planting or plant the cluster and thin the seedlings later. It is usually easier to thin the seedlings, and you can add the thinnings to salads or other dishes.

Once established, Swiss chard requires little more than regular watering, which is especially important in hot weather, when growing slows and the leaves can become bitter. The plants can handle a few light frosts, so with some protection, even gardeners in cold climates can harvest into midwinter.

Swiss chard leaves fall prey to many insect pests. Most insect damage can be avoided by using row covers. Slugs and snails can be collected and destroyed with traps. Aphids may congregate, but a good hosing off, several

Exposure:
Full sun to partial shade

Ideal soil temperature:
55–75°F (13–24°C)

Seed Planting depth:
1/2–1 in. (1–2.5 cm)

Days to germination:
7–14 days

Spacing:
4–6 in. (10–15 cm)

Days to maturity:
55–60 days

'Rainbow Chard' is one of the earliest greens to emerge in the spring, and it keeps growing well into the fall or winter.

days in a row, should thwart them. The worst pest is beet leaf miner, fly larvae that tunnel throughout the leaves. Remove any infested leaves and protect the plants with row covers. A natural product called Spinosad shows great promise for leaf miner control.

How to harvest Harvest individual outer leaves or slice off the whole plant about 2 in. (5 cm) from the ground. The plant will regrow more leaves for later harvests. Leaves of about 10 in. (25.5 cm) or shorter will have the best flavor.

Others to try Try 'Flamingo Pink' (with neon pink stems), 'Oriole Orange', and 'Rhubarb Chard' (bright red stems).

CORN

Would it be summer without sweet corn? No mere ornamental grass, corn blurs the lines between vegetable and grain. One small square of soil is enough growing space to treat yourself to freshly picked corn that is so tender it needs no cooking. Or perhaps your taste runs to the comforting crunch of popped corn. Those tiny ears are even easier to grow.

Corn is native to the Americas and developed from a wild grass at least 7000 years ago. The term "corn" was used in Europe to refer to any grain that was an important food crop for a region, whether it be wheat, oats, or barley. The early settlers used the designation "Indian corn" for the grain grown in the new world. The Pilgrims embraced corn after much of the seed they brought with them rotted during their long voyage.

Great heirloom corn combines the best of the grainy, cereal essence of corn with a juicy, sugary sweetness. Nathan Stowell bred his namesake corn in the 1840s with a little extra bonus: an unusual storage quality from a technique he adopted from the Iroquois. 'Stowell's Evergreen' is a winter corn that can be pulled out of the ground, stalk and all, and hung upside down in a cool, dry storage area to finish ripening and be harvested later. This clever technique allows you to extend the fresh corn season well into winter.

'Stowell's Evergreen' is still a popular garden corn, although poor Nathan would probably derive no pleasure from it. Story has it that he sold a friend a couple of ears for $4 and the friend then sold the seed to a seed company for $20,000.

Exposure:
Full sun

Ideal soil temperature:
75-85°F (24-30°C)

Planting depth:
1 in. (2.5 cm)

Spacing:
2-4 in. (5-10 cm), in blocks of at least three rows

Days to germination:
7-10 days

Days to maturity:
95 days

Corn 'Stowell's Evergreen'
Sweet corn
Zea mays

Flavor 'Stowell's Evergreen' has just enough sweetness, unlike most modern sweet corns, and a rich, grainy flavor that balances its natural sugars. You can savor the essential character of corn when you pick and eat it fresh, before the sugars turn to starch.

Growing notes Direct sow this late-season variety in late spring to early summer. Gardeners in frost-free climates can sow seeds in late summer for a fall harvest, but germination is lower in summer heat.

'Stowell's Evergreen' is a sweet corn that stores well, if you can wait to eat it.

Corn is pollinated by the wind, so grow several plants close together to ensure that enough pollination occurs for the ears to fill with kernels. If you have the space, thin plants to 12 in. (30.5 cm) apart, with at least three rows spaced 24 in. (61 cm) apart. Even in a small space, three rows with plants thinned to 6–8 in. (15–20 cm) apart should provide enough corn for a few meals.

Corn is a heavy feeder. Add some alfalfa or soybean meal to boost the soil's nitrogen level. Then give the plants a dose of balanced fertilizer when the tassels first appear.

How to harvest You can try Nathan Stowell's cold storage method or simply pick the ears to enjoy fresh. The ears should feel full, although they do not always fill to the end. Piercing a kernel with your fingernail will produce a milky juice when the corn is ready.

Tip Raccoons are infamous for stealing corn just before it is ripe. Hang old CDs around your corn; they will reflect the sunlight and deter the animals, at least during the day. A big dog is your best defense against these thieves.

Others to try 'Art Verrell's White' is a short-season white sweet corn well suited to gardens in cooler climates. 'Country Gentleman' is a great canning and creaming corn. 'Golden Bantam Improved' is the standard for yellow sweet corn.

'Lady Finger' sounds delicate, and it is. This Amish heirloom is not only light and fluffy, it is considered hulless: you get the crunchy popcorn flavor, and no hulls get stuck in your teeth. Of course, the kernals are not completely hulless or they would fall apart. But the hull is small enough to crumble and virtually disappear as it pops. These slender 6 in. (15 cm) ears hold golden yellow kernels with a touch or two of reds and purples splashed throughout.

Popcorn is one of the oldest varieties of corn, dating back some 5000 years. Native Americans discovered that it could be popped and eaten while still on the cob. The early colonists popped corn for breakfast, topped with cream and sugar. 'Lady Finger' is too delicate to drown in cream, but a little melted butter would not be out of order.

Exposure:
Full sun

Ideal soil temperature:
75–85°F (24–30°C)

Planting depth:
1 in. (2.5 cm)

Spacing:
3–6 in. (7.5–15 cm)

Days to germination:
7–10 days

Days to maturity:
95–105 days

Popcorn 'Lady Finger'
Zea mays var. praecox

Flavor 'Lady Finger' brings a full, fresh, nutty corn taste, with the delicate crunch of a tender kernel, free of chewy hull pieces. Eating it is like eating a fluffy, crisp, saltless corn chip. You will want more than a single handful.

Growing notes Direct sow after the soil has dried and all

Delicate 'Lady Finger' is a hulless, tiny popcorn.

danger of frost has passed. You can start corn indoors, but use peat or paper pots so you do not disturb the tap roots when transplanting. Gardeners in frost-free climates can sow 'Lady Finger' in midsummer for a fall harvest.

Start with a rich soil, amended with lots of compost. Sow seeds after the ground has dried out and warmed to about 70–75°F (21–24°C). Corn is pollinated by the wind, and several plants can be clustered together to ensure enough pollination for the ears to fill with kernels. To enjoy a few good-sized servings of popcorn, plant at least a 4-by-4 ft. (1.2-by-1.2 m) block divided into 1 ft. (0.3 m) squares. Start with 2–4 seeds in each square and thin seedlings to the strongest plants.

Keep the seeds moist until they germinate. Corn is a heavy feeder and requires a lot of water while the ears are forming. Give the plants a dose of balanced fertilizer when the tassels first appear.

When to harvest Harvest the ears when the stalks and husks turn brown and dry and the ker-nels feel hard. To harvest, twist the ears off the stalks. If the season is damp or frost threatens, you might need to harvest earlier and allow the ears to finish drying off the plant.

After harvest, popcorn requires another month of drying, or curing. Spread the ears on screens or place them in mesh bags and allow them to dry for 4–6 weeks in a warm, dry spot with plenty of air circulation. When the corn is thoroughly dry, store it in airtight containers, either as whole ears or just kernels. Remove kernels by twisting the ears with both hands. Gloves are advised.

Tip 'Lady Finger' popcorn makes a festive fall decoration as it dries and grows ready for popping.

Others to try 'Japanese Hulless' has tiny, tender, white kernels. 'Little Indian' has multicolored kernels that pop to a tender crunch. 'Tom Thumb' has sweet, small ears of yellow hulless popcorn. It was developed by plant breeder and University of New Hampshire botanist Professor E. M. Meader.

CUCUMBERS

Cucumbers are so juicy that eating one is like taking a cool sip of water. Few vegetables are more refreshing on a hot summer day. Why limit yourself to the long slicer? Cucumbers can be found in a variety of shapes and colors that dress up a dish or literally become the dish itself. Some cucumbers have a natural tangy, sour flavor, such as 'Mexican Sour Gherkin', that taste as though they have already been pickled. Now that's pretty cool.

Cucumbers are native to northern India and are known to have been grown more than 3000 years ago. The European explorers brought cucumbers to the New World.

'Boston Pickling' may have made a name for itself because of its use in brine, but this is one of the best fresh eating cucumbers you will ever taste. A cucumber has to be firm and crunchy to make a great pickle, and 'Boston Pickling' is legendary. Short and stout, with juicy, crisp flesh and a small amount of pulp and seed, this cuke will be eaten as fast as it keeps producing.

A favorite since at least 1880, it goes by many names, including the descriptive 'Green Prolific'. This is a dependable, versatile, and never bitter cucumber.

Exposure: Full sun

Ideal soil temperature: 70–85°F (21–30°C)

Planting depth: 1/2–1 in. (1–2.5 cm)

Days to germination: 7–10 days

Spacing: 6–8 in. (15–20 cm)

Days to maturity: 50–60 days

Cucumber 'Boston Pickling'

'Green Prolific'
Cucumis sativus

Flavor 'Boston Pickling' cucumbers are refreshingly cool with an almost grassy brightness. Because they are quick growers and are harvested small, the skin is thin and tender and the juicy flesh stays sweet, firm, and crunchy. This is an ideal all-around cucumber for pickles, salads, and crudité.

Growing notes Direct sow after all danger of frost has passed. You can also start seeds indoors 3–6 weeks before transplanting out. A second sowing can be done in midsummer for a later harvest. Gardeners in frost-free climates can sow seed in late summer to early fall.

'Boston Pickling' cucumbers are small, but the vine can easily reach 6–8 ft. (2–2.4 m). You can let it sprawl on the ground, but the cucumbers are easier to find if the plants are trellised. Keeping the fruits off the ground will also help prevent disease problems and rotting and will keep them away from hungry animals.

To help the cukes stay juicy and sweet, water the plants regularly and mulch the roots to keep them cool. Plants will stay healthy and productive if you give them a dose of fertilizer or side dress with compost every 3–4 weeks.

Although you need to watch for powdery mildew and cucumber beetles, 'Boston Pickling' is resistant to mosaic virus and cucumber scale.

How to harvest Harvest when the cucumbers are 3–6 in (7.5–15 cm). Stay on top of the harvest, because 'Boston Pickling' is prolific, and the more cukes you harvest, the more you will get.

Tip A large tomato cage makes an easy, space-conserving trellis for a cucumber. Place the cage over the cucumber at planting time, and train the vine around the outside of the cage as it grows.

Others to try 'A & C Pickling' is an easy growing, full-sized cucumber that is tasty fresh or pickled. 'Early Russian' produces short, thin fruits that are sweet and early maturing. 'Japanese Climbing' produces long, bumpy fruits that make great pickles when young and are tasty fresh cukes when mature.

'Crystal Apple' produces palm-size, creamy white cucumbers with glowing green stripes running from top to bottom. They are so cool, fresh, and juicy that they can be eaten like apples. These cucumbers grow to apple size as they mature, but I would recommend that you not wait to harvest them at maturity. Like most cucumbers, they get seedier as they age.

This cuke comes from New Zealand, where it is still popular. New Zealand growers suggest it be harvested when it is the size of a kiwi, a good size for slicing. It has proven to be surprisingly drought-resistant and a long season producer. Delicate looking vegetables always seem to have something to prove.

PROLIFIC

Exposure:
Full sun

Ideal soil temperature:
70–85°F (21–30°C)

Planting depth:
1/2–1 in. (1–2.5 cm)

Days to germination:
7–10 days

Spacing:
12 in. (30.5 cm)

Days to maturity:
65 days

Cucumber 'Crystal Apple'
Cucumis sativus

Flavor 'Crystal Apple' has a tender skin an apple would envy. It needs no peeling; in fact, the skin adds to the juicy, crisp texture. Despite its name, do not expect it to taste of apples, unless you are extremely impressionable. Picked young and firm, it has a clean, grassy flavor. Refreshing is the best adjective for this cuke.

Growing notes Cucumbers are best direct seeded in the garden after all danger of frost has passed. The plants tend to exhaust themselves after producing fruits, and reseeding every 3–4 weeks will extend the harvest. Gardeners in frost-free climates can begin seeding in midsummer for a fall and winter harvest.

The vigorous vine has good disease resistance and produces fruit for a long time. As with most cucumbers, it has shallow roots and requires regular water, especially while the fruits are forming, which is most of the time. Mulching around the plants will help the soil retain moisture.

A cucumber this prolific needs some supplemental feeding midsummer; side dress with compost or manure. Whether or not you have the space to let them sprawl, growing them vertically on some type of trellis is advised. This will keep the pale fruits clean and round.

How to harvest Unless you are saving seed, harvest fruit before they reach apple size and start to get overly ripe: 2–3 in. (5–7.5 cm) in diameter is about right. Pick them before the green markings start to turn yellow.

Both the fruits and the vines can be prickly. Sensitive gardeners should wear gloves while harvesting. You can twist cucumbers off their vines, but they can be stubborn and the vine can come off with the fruit; better to cut the cukes off the vines.

Others to try 'Miniature White' is small and sweet, and the skin is good to eat. 'Richmond Green Apple' is round and green with crispy white flesh. 'White Wonder' is a juicy and lovely elongated ivory cucumber.

Everything about this cucumber is delicate, but impressive. The vines appear so deceptively dainty you might think they will disintegrate on touch. The flowers resemble buttercups, and the tangy little fruits look like miniature watermelons. In fact, although 'Mexican Sour Gherkin' is in the same family as cucumbers and is generally used as a pickled cucumber, it is more accurately categorized as a melon.

Its Spanish name, sandita, means little watermelon. You may also see 'Mexican Sour Gherkin' listed as Guatemalan cucumber, mouse melons, or miniature watermelons. It is a native of Mexico and Central America, where it has been grown since before Europeans came calling. This is a very old heirloom that is starting to regain popularity. If only we had more recipes for using it.

Cucumber 'Mexican Sour Gherkin'

Mexican miniature watermelon, mouse melon
Melothria scabra

Flavor The white flesh has a fresh cucumber flavor and crunch, but the skin surprises you with an acidic tartness similar to a tomato. It is a refreshing combination, similar to that of a young plum. With their characteristic tangy tartness, they taste already pickled and are ready to dress up salads. The small, 1–2 in. (2.5–5 cm) fruits are almost bite-sized and can also be tossed in stir-fries, used as appetizers, or added to salsas.

Growing notes Direct seed in the garden after all danger of frost has passed, or get a head start indoors, 3–6 weeks before transplanting out. Plant a second batch in midsummer for a later harvest. Gardeners in frost-free climates can start seed in late summer to early fall.

Start with warm, dry, rich soil and feed plants monthly with a dose of balanced fertilizer or a side dressing of compost. 'Mexican Sour Gherkin' is grown and tended like a cucumber. It is a bit of a slow starter, but it makes up for it by producing prolific 10 ft. (3 m) vines. A trellis is strongly recommended or the vines become a knotted pile. It is tolerant of cool weather and should keep producing until frost. Be sure to provide plenty of water, especially during dry spells.

How to harvest The tiny fruits would be difficult to find buried in the foliage, but they make it easy for you by falling off the vine when they are fully ripe. Unlike cucumbers in the genus *Cucumis*, which need to be harvested when slightly immature, 'Mexican Sour Gherkin' stays juicy and flavorful when allowed to ripen. Fruits less than 1 in. (2.5 cm) in diameter are the most tender for eating fresh.

Tip Many gardeners have been disappointed in the productivity of their cucumbers, because plants in the cucurbit family have separate male and female flowers and both must be present at the same time for the

Exposure:
Full sun

Ideal soil temperature:
75–85°F (24–30°C)

Planting depth:
1/2–1 in. (1–2.5 cm)

Days to germination:
7–10 days

Spacing:
12 in. (30.5 cm)

Days to maturity:
65–75 days

A crisp, juicy mouse melon has a faintly acidic tomato flavor.

plants to be pollinated and form fruits. *Cucumis* species plants set male flowers first, and it can take weeks for the fruit-forming female flowers to appear. But 'Mexican Sour Gherkins' begin setting their female flower buds first, ensuring a plentiful harvest.

Others to try These are not as tangy as 'Mexican Sour Gherkin', but they make great pickles: Two good pickling cukes (*Cucumis sativus*) are 'Home-made Pickles', with short, white fruit and dense, crisp flesh, and 'Parisian Pickling', also known as 'Improved Bourbonne', a crunchy French heirloom that is perfect for small gherkins. Also try West Indian or burr gherkins (*Cucumis anguria*), a small, spiny, and sweet pickling cucumber.

'Poona Kheera' can disguise itself as a potato, but one bite will tell you otherwise. This little bundle is about as crisp and juicy as a cucumber can get. The short fruits literally snap in half. Morphing from pale green to yellow to a russet brown, these Indian cucumbers stay tasty throughout their growth cycle.

The cucumbers' unusual color is considered a tough sell to North Americans, and they are often marketed while young and pale green. Do not be put off by the color of the skin, however, because they are at their crispy sweetest just as they are peaking from yellow to russet. Southeast Asian cucumbers are often crosses between cucumbers and melons, which gives them their distinct luscious flavor and texture.

Cucumber 'Poona Kheera'

Cucumis sativus

Flavor The most imposing quality of 'Poona Kheera' is its crispness. Its firm texture allows you to savor the flavors longer. The level of sweetness builds as the fruits mature, along with the size of the seeds. It is most flavorful and tender when it is starting to shade over in yellow. It has the pungency of a cucumber, with a refreshing brace.

Growing notes Direct seed in the garden after all danger of frost has passed, or get a head start indoors, 3–6 weeks before transplanting out. Try a second seeding in midsummer for a later harvest. Gardeners in frost-free climates can start seed in late summer to early fall.

'Poona Kheera' is a quick grower that slows down a bit in extreme heat. It is wise to start a second planting in midsummer if you want to keep harvesting into the fall. Start with a rich soil that has warmed and dried out a bit.

The vines do not get particularly heavy, and the fruits are harvested small, so the plants do not need to be trellised unless you need to save room. Its unusual color makes the cucumbers easier to find under the foliage than traditional green cukes.

Add a dose of fertilizer or a side dressing of compost every 3–4 weeks. To help them maintain their cool crispness, give these cukes plenty of regular water, especially during dry spells.

How to harvest Everyone has his own preference for the best time to harvest 'Poona Kheera'. The best idea is to experiment the first time you grow them, and determine at what stage you like them best. The fruits can be so crisp they are brittle, so use a knife or pruners to cut them from the vine.

Tip When cucumbers start to bulge in the middle, they are past prime and filling with seeds.

Others to try 'Chinese Yellow' has blunt, yellow to orange fruits that make great pickles. 'Edmondson' has sweet and tender cream-colored fruits. 'Mini White' grows sweet 3 in. (7.5 cm) fruits on a short, well-behaved vine.

Exposure:
Full sun

Ideal soil temperature:
75–85°F (24–30°C)

Planting depth:
1/2–1 in. (1–2.5 cm)

Days to germination:
7–10 days

Spacing:
12 in. (30.5 cm)

Days to maturity:
55–65 days

(Above) 'Poona Kheera' can be enjoyed before it matures to a russet brown color. (Below) This is one time when you can let your cucumbers turn yellow. 'True Lemon' is at its best when it is just starting to shine.

'True Lemon' cucumbers resemble pale yellow lemons, but their flavor does not cause any puckering. They are crisp and juicy, with a refreshing brightness that is almost as surprising in a cucumber as it is in a lemon.

This cucumber is an old heirloom, said to have been introduced in 1894, but some form or another has probably been grown for a lot longer. A lemon cucumber popular in India, dosakaya, has long been used in pickles and chutneys.

Cucumber 'True Lemon'
Cucumis sativus

Flavor These cucumbers are mildly sweet, very juicy, and extremely tender. While they are pale yellow and about the size of a jumbo lemon, they need no peeling. They will turn a brighter yellow as they age, but they also get seedier, with scratchy black spines and a bitter taste more befitting their name. This great general purpose cucumber makes terrific pickles and great snacks, but you can also cut them in half and scoop out the middle to create edible bowls. Add a scoop of chicken salad or tabouli and dig in.

Growing notes Direct seed in the garden 3–4 weeks after all danger of frost has passed. Planting every 4–5 weeks into midsummer will ensure healthy vines that keep producing. Gardeners in frost-free climates can sow seeds in midsummer to late summer for a fall harvest.

Cucumbers have shallow roots and depend on regular watering, which is vital while the fruits are forming. Mulching around the plants will help the soil retain moisture. Side dressing with compost or manure in midsummer will help them keep producing, but succession planting it the best way to ensure cucumber harvests all summer long.

Although 'True Lemon' cucumbers are not as heavy as larger varieties, the vines can sprawl. Trellising saves space in the garden, makes harvesting easier, and keeps the fruits lifted off the ground and away from animals. It also allows the fruits to form as nice round balls, without the bruising and marking that can occur when they lie on the ground.

How to harvest Harvest when the cukes are about 2–3 in. (5–7.5 cm) in diameter. Both the fruits and the vines can be prickly, so you might want to wear gloves while harvesting. You can twist cucumbers off their vines, but they can pull the vine with them. Better to cut the fruits off the vines.

Tip For almost instant pickles, slice the cucumber and pour a hot vinegar pickling mix over them. Let them cool in the refrigerator and they are ready to serve.

Others to try 'Boothby's Blonde' is a small, pale yellow, Maine family heirloom the size of a fat cigar, with a sweet taste and creamy texture. 'Dragon's Egg' is an egg-shaped, creamy white, nonbitter, and prolific cuke.

Exposure:
Full sun

Ideal soil temperature:
75-85°F (24-30°C)

Planting depth:
1-1½ in. (2.5-4 cm)

Days to germination:
7-10 days

Spacing:
12 in. (30.5 cm)

Days to maturity:
60-75 days

EGGPLANTS

The eggplant is like no other vegetable. Distinctive in appearance and flavor, eggplants turn an ordinary meal into a captivating ethnic dish. These beautiful, glossy fruits can be plump and purple or pretty much any color and shape you could want. Eggplant is a perfect addition to whatever you are planning to whip up, from rustic ratatouille, to an elegant grilled pizza.

'Japanese White Egg' offers a tiny burst of intense eggplant flavor in a surprising package. The plants' kiwi-sized white fruits are shaped like perfect tiny eggs. We should not be so surprised, however, because the first varieties introduced in England were egg-shaped, and the English gave them their descriptive name.

Thomas Jefferson brought the first eggplant seed to the United States in 1806, and seeds are still sold at Monticello. The plants produced small, white, egg-shaped fruit. It was considered more of a novelty at the time, and even the aubergine eggplants were grown mostly as ornamental plants until the mid-1900s. American gardeners have since fallen in love with these prolific plants, and we appreciate the rich and flavorful fruits.

Exposure: Full sun

Ideal soil temperature: 75–85°F (24–30°C)

Planting depth: 1/4–1/2 in. (0.6–1 cm)

Days to germination: 7–14 days

Spacing: 18–24 in. (45.5–61 cm)

Days to maturity: 65–75 days

You might grow 'Japanese White Egg' for the novelty, but you will soon learn to love its versatility in the kitchen.

Eggplant 'Japanese White Egg'

Solanum melongena var. *esculentum*

Flavor 'Japanese White Egg' offers a dense flesh with a full, rich eggplant earthiness reduced to a compact size. Although these eggplants have a lot of seeds, they are small and do not add bitterness. The skin can get a bit thick as the fruits mature; harvest them young and tender or use the larger fruits in baking and frying when you what them to hold their shape. They make a nice pickle, too.

Growing notes Because eggplants require a long growing season, gardeners in cool climates should start seed indoors 8–10 weeks before the last expected frost and transplant when nighttime temperatures are reliably warm. Eggplant seeds are slow to germinate. Bottom-heating the seed flats will help move them along. Gardeners in Zones 9 and warmer can plant in midsummer and again in fall.

This is one of the least fussy eggplant varieties to grow. It stands upright even with a dozen dangling fruits, barely bending with the weight. All eggplants thrive in heat and need plenty of sun and regular water. Although it is a heat lover, it can still suffer stress during long, oppressive heat waves. Give your plants a little shade, or at the very least water them daily to prevent wilting. Stressed plants are magnets for feeding insects.

Feed after the fruits begin to form and then monthly after that. 'Japanese White Egg' has good disease resistance and is tolerant of extreme weather.

How to harvest Harvest fruit any time after they reach about 1/2 in. (1 cm) in diameter. They will be more tender when they are small. Fully ripe fruits turn yellow and have tougher skins.

Others to try 'Casper' is a full-sized white eggplant with an earthy, mushroom flavor. 'Chinese White Sword' is an Asian eggplant that produces long, thin, tender fruits. 'Little Spooky' is a Japanese variety with large tasty fruits and is said to scare away evil spirits.

Mild, tender, and a virtual sponge for other flavors, 'Ping Tung Long' is among the best of the Asian eggplants. It holds its texture when cooked, staying firm and pleasant. It shares the purple coloring of Italian eggplants but offers a tender, glossy skin that needs no peeling.

Named after Pingtung, Taiwan, it has the small, sparse seeds and absence of bitterness common in Asian eggplants. Starting fairly early in the season, the compact plants virtually hunch over with heavy crops of perfect, shiny, bright lavender fruits. It matures quickly and keeps producing throughout the summer.

Eggplant 'Ping Tung Long'

Solanum melongena var. *esculentum*

Flavor 'Ping Tung Long' has a gentle, herbal sweetness, with little aftertaste. The firm but creamy texture is perfect for chopping and adding to stir-fries and curries, or for slicing and roasting. Use the fruits within a few days of harvesting to enjoy the delicate and sumptuous flavor.

Growing notes In cool climates, start seeds indoors 8–10 weeks before your last expected frost. Eggplant seeds are slow to germinate. Adding bottom heat to seed flats will help move them along. Gardeners in Zones 9 and warmer can plant in midsummer and again in fall. Fall planting is not possible in cool climates.

'Ping Tung Long' starts producing early and tends to be more tolerant of cool weather than most heat-loving eggplants, although it still resents a chill. Wait until the nights are reliably warm before setting plants outdoors.

This eggplant is not particular about soil, as long as it is warm. Sun and regular watering will keep the plants healthy and the fruits plump. Feed after the fruits begin to form. The plants are somewhat compact, but the high yields can cause them to topple over. Staking the plants keeps the fruits from dusting the ground and also helps them grow straighter.

'Ping Tung Long' has good disease resistance and is tolerant of extreme weather, but all eggplants are prone to the same diseases as their cousins, tomatoes and peppers. Flea beetles are a fact of life with eggplants. Healthy plants will generally outgrow the affects of flea beetles.

How to harvest Start to harvest 'Ping Tung Long' as small eggplants or let them grow to about 6–8 in. (15–20 cm). They are glossy and firm to the touch when ripe. Cut them from the stem, leaving the green calyx attached to the fruit. The stems are scratchy, so watch your hands.

Tip Growing eggplants in pots will keep the soil and roots warm and produce more fruits in cool climates.

Others to try 'Fengyuan Purple' is one of the thinnest, longest Asian eggplants and is good for grilling. 'Ma-Zu' is thin and tender, and great for stir-fries. 'Thai Long Green' has a wonderful texture and sweetness.

Exposure:
Full sun

Ideal soil temperature:
75–85°F (24–30°C)

Planting depth:
1/4–1/2 in. (0.6–1 cm)

Days to germination:
7–14 days

Spacing:
18–24 in. (45.5–61 cm)

Days to maturity:
65–75 days

'Ping Tung' is a nice shape for grilling long slices.

'Rosa Bianca' was one of the first of the stunning, thin-skinned heirloom eggplants to find favor with gardeners and cooks. The texture is often described as creamy, but in this case, it is more like melt-in-your-mouth velvetiness. This is not the giant purple orb most of us expect from an eggplant. It is much smaller than traditional eggplants, and its color is a violet-purple shading over a white base. The name means white rose, hinting at the esthetic value of eggplants in the garden. Eat this eggplant while it is plump and fresh, with skin so tender that no peeling is required.

Exposure:
Full sun

Ideal soil temperature:
75–85°F (24–30°C)

Planting depth:
1/8–1/4 in. (0.3–0.6 cm)

Days to germination:
7–14 days

Spacing:
18–24 in. (45.5–61 cm)

Days to maturity:
80–90 days

(Above) Eggplant 'Rosa Bianca' starts to fade in color as it matures, a sure sign you have waited too long to harvest. (Below) 'Turkish Orange' will turn a brilliant, glossy orange, but it tastes best when it is just starting to change color.

Eggplant 'Rosa Bianca'

Solanum melongena var. *esculentum*

Flavor Beauty aside, 'Rosa Bianca' also boasts tender skin and dense flesh that is firm and smooth, rather than the spongy consistency we expect from a commercially grown eggplant. With a more subtle flavor than traditional eggplants, it offers sweetness, especially when roasted. The fruits are bitter only if they are not picked at glossy perfection.

Growing notes In cool climates, start seeds indoors 8–10 weeks before your last expected frost date. Eggplant seeds are slow to germinate. Bottom heating seed flats will help move them along. Gardeners in Zones 9 and warmer can plant in midsummer and again in fall. Fall planting is not possible in cool climates.

Heat and sun are essential to producing eggplants. They will not set fruit in temperatures lower than 70°F (21°C). In cooler climates, growing the plants in pots will help keep the soil warm and encourage production. Because they are soaking up the hot sun, regular watering is essential.

Help them along with a side dressing of compost or manure after the fruits start forming.

'Rosa Bianca' can be a large, bushy plant. Although only two or three fruits will appear on the plant at one time, they can become heavy. Stake them to keep them upright and allow the sun to reach the center of the plants.

Eggplants are prone to the same diseases as tomatoes and peppers. Try to rotate them in your garden in subsequent years to avoid problems. Flea beetles are the most serious pests, although the damage they cause is mostly cosmetic.

How to harvest Start checking your eggplants as they approach mature size, 4–6 in. (10–15 cm). They should be glossy and firm to the touch. Cut them from the stem and leave the green calyx attached. The stems are scratchy, so watch your hands.

Tip Planting your eggplants near onions will keep most flea beetles at bay.

Others to try 'Listada de Gandia' is sweet and tender, and it grills nicely. 'Ronde de Valence' offers rich flavor and a round shape; it is great for stuffing. 'Round Mauve' is small, with few seeds and no trace of bitterness.

'Turkish Orange' is a whole new world of eggplants. Its neon orange, glowing skin is so glossy it looks polished. It is also very misunderstood. Although it is named for its color, it should be eaten just before it turns orange or it will taste very bitter. When harvested at its best, it has soft, juicy flesh and a tangy, sweet fragrance.

'Turkish Orange' does indeed hail from Turkey, where is has been grown since the 15th century. It is also popular in Italy, where they know a thing or two about both tomatoes and eggplants. In fact, this eggplant is often used like a tomato in recipes.

COLORFUL

Eggplant 'Turkish Orange'
Scarlet eggplant
Solanum aethiopicum

Flavor If you close your eyes and forget about its appearance, 'Turkish Orange' offers some of the earthy eggplant flavor of traditional eggplants. But a sparkling zestiness is layered on top of that earthy note. This eggplant is more refreshing than the usual eggplant and less acidic flavored than tomatoes. It makes a wonderful addition to curries, it can be stuffed, or it can be simply sliced in half and grilled until warm.

Growing notes Start seeds indoors 8–10 weeks before your last expected frost date, and transplant outdoors after the soil has warmed to about 70–75°F (21–24°C). Gardeners in Zones 9 and warmer can plant in midsummer and again in fall.

'Turkish Orange' might not look much like an eggplant, but it grows under much of the same conditions and care as any other eggplant variety. It prefers a moderately rich soil that is well-draining. It also needs sun, regular watering, and a monthly dose of fertilizer, fish emulsion, or compost tea to grow and produce healthy, succulent fruits all season long. The plants grow 2–3 ft. (0.6–0.9 m) tall with a dozen or more fruits per plant. Staking is not required if you harvest frequently.

How to harvest The peak of flavor occurs when the fruits still show some green striping, before they turn completely orange. They should be glossy and just beginning to soften. Although visually stunning when fully orange, they are also stunningly bitter in taste. Spare yourself.

Tip If a few fruits ripen before you can harvest them, pick them anyway to keep the plant producing. They can always be used in flower arrangements.

Others to try 'Brazilian Oval Orange' has small oval fruits with a flavor that falls between mild Italian eggplants and bitter Asian varieties. 'Ruffled Red' is a similar eggplant from Thailand, with a spicy taste suited to Asian cuisines. 'Striped Toga' produces a small, egg-shaped, orange fruit with a more traditional eggplant flavor.

Exposure:
Full sun

Ideal soil temperature:
75–85°F (24–30°C)

Planting depth:
1/4–1/2 in. (0.6–1 cm)

Days to germination:
5–14 days

Spacing:
18–24 in. (45.5–61 cm)

Days to maturity:
75–90 days

LEGUMES

Beautiful, versatile, and scrumptiously good for you, beans are treasures that were once used as currency. More than 1000 different kinds of beans are incredibly undemanding, eager growers. Whether snapped or shelled, green, speckled, striped, or winged, beans jazz up any meal.

Beans have been grown since at least 8000 BC. Bean seeds made their way from Central and South America to Europe sometime around 1500. Their small size made it easy for large quantities and varieties to travel easily. By the mid-1900s, the Netherlands had acquired an impressive collection of 1500 distinct varieties. Beans can be truly dazzling, visually and in terms of flavor. They are served as delicate French haricot verts, hearty Boston baked beans, and spicy Cuban black beans, to name a few.

Fresh peas signal the start of the growing season. The first pods rarely make it to the table; in their tidy single-serving packages, you can just pop open a pod and taste them while working in the garden. Tangy and sweet, peas have a fresh taste that alerts you to the promise of things to come. Their brief growing season connects us to the seasonality of foods, a notion all but forgotten in the age of globally grown and shipped produce. The delight of a garden fresh pea is a treat a child will remember well into adulthood.

Peas are one of the earliest cultivated plants, too. They were grown in the Stone Age, although fresh, shelled peas did not become popular until the 1600s. They are not the luxury they were back then, except when the growing season ends and we have to wait another year.

'Cannellini' beans are the famed white shelling beans of Tuscan minestrone, and their widespread use is one of the reasons Tuscans were called mangiafagioli, or bean eaters. These beans have been available in North America since before 1900, but they are not often grown in home gardens and seed is difficult to find. You will almost never find these beans sold fresh in supermarkets, because they are easier to shell when dried.

The shape of these beans is a tip off that they are related to kidney beans; in fact, they are frequently called white kidney beans. Seek out the true 'Cannellini' beans to savor their delicate creaminess. Do not confuse them with the smaller great northern or navy beans or the larger, bolder, runner cannellini pole bean.

Exposure:
Full sun

Ideal soil temperature:
75-80°F (24-27°C)

Planting depth:
1–1½ in. (2.5-4 cm)

Spacing:
4-6 in. (10–15 cm)

Days to germination:
7–14 days

Days to maturity:
70–80 days shelling, 100 days dry

When cooked, the lovely 'Cannellini' beans plump up into hearty, meaty beans.

Bean 'Cannellini'
Cannellone bean
Phaseolus vulgaris

Flavor 'Cannellini' beans' tender skin holds up well in cooking and protects the smooth, melt-in-your-mouth texture of the interior. When eaten plain, they have a mild, nutty, earthy flavor, but they have a wonderful ability to pick up other flavors and seasonings, whether in a simple sauté with a sprinkle of thyme or simmered in the strong flavors of tomatoes and garlic.

Growing notes Direct seed in the spring, after the soil has warmed to about 50°F (10°C). They can be succession sown every 2–3 weeks into midsummer to extend the harvest. 'Cannellini' beans prefer cooler temperatures and do better if gardeners in Zones 8 and warmer wait to sow in late summer or fall. You can start beans indoors in peat or paper pots if animals or insects are a problem for tender seedlings.

These bush beans grow to about 3 ft. (0.9 m) tall, but if the pods are left on the plants until they are dry, you may want to give the plants some support, because they can become floppy. They also need lots of sunlight and room for air circulation. They are light feeders, so planting the seeds in rich soil with plenty of organic matter is all they will need.

How to harvest Harvest 'Cannellini' beans for fresh eating after the pods fill out. Mature pods will be about 4 in. (10 cm) long. If you prefer to dry the beans for storing, let the pods dry thoroughly on the plant and then harvest and shell them. As the walls of the pods dry, the seeds loosen and are easier to remove.

Tip Cooking the beans slowly helps retain their texture: simmer dry beans for 2–3 hours until they are tender.

Others to try 'Borlotto' and cranberry beans are speckled red beans, slightly sweeter than pinto beans and popular in Europe and New England. Garbanzo beans (*Cicer arietinum*, also called chickpeas) are round with a buttery, peanut flavor, popular in Mediterranean and Indian dishes. Great northern beans cook quickly and are often used for baked beans.

'Chinese Red Noodle' beans are not actually beans; they are in the legume (Fabaceae) family, however. A relative of cowpeas, they deliver a distinct pungent flavor in the versatile package of a long, thin, quick-cooking pod.

It would be easy to dismiss 'Chinese Red Noodle' as an ornamental plant. The lavender blossoms grow into a pair of burgundy-colored ribbons that stretch into dangling "noodles" within days. But you will be rewarded at the table if you resist the urge to leave the pods on the plant to admire. Pick them and more will follow

Bean 'Chinese Red Noodle'

Chinese long beans, red asparagus beans
Vigna unguiculata subsp. *sesquipedalis*

Flavor 'Chinese Red Noodle' beans offer the snap of green beans and a whisper of legume flavor, with layers of the earthier flavors of asparagus, mushrooms, and their cowpea cousins. They tend to be drier and nuttier than cowpeas, and diners will wonder what seasonings you added to the meal to achieve such a distinctive taste. If you are familiar with green asparagus beans, 'Chinese Red Noodle' has a slightly firmer texture and a more robust taste. They are great in stir-fries and pair nicely with peppers, whether sweet or hot.

Growing notes Direct seed in the garden in moderately rich soil, 2–3 weeks after frost danger has passed. Reseed in 3–4 weeks to prolong the harvest. If you garden in a frost-free climate, you can continue seeding into fall.

Seeds will rot in cold, wet soil, so wait until the soil has dried enough to crumble in your hand before planting. These tall pole beans need some sturdy support over which to climb and sprawl. Install a trellis or poles in the ground when you seed so you can begin training the vines early. The vines will take a while to become established, but once they start to flower, they will produce heavily.

These beans are not heavy feeders and, as legumes, they fix their own nitrogen, so supplemental fertilizer is not required. However, they will stop flowering in dry spells without extra water.

How to harvest Bean pods will grow to 20 in. (51 cm) or more, but they are best used when 12–15 in. (30.5–38 cm) long, before the seed shapes begin to show through the pods. Carefully snap or cut the beans from the vines to avoiding pulling the vines down with them.

Tip Do not be alarmed if you see ants crawling over the blossoms and beans. They are attracted to the sap and will not hurt the beans. They may bite you, though, so shake them off before you harvest.

Others to try 'Chinese Green Noodle' has a milder flavor and is extremely productive. 'Chinese Mosaic' has 12–18 in. (30.5–45.5 cm) lavender beans with a nutty, savory flavor. 'Red Stripe Seed' has a rich, meaty bean flavor.

Exposure:
Full sun

Ideal soil temperature:
65–80°F (18–27°C)

Planting depth:
1–2 in. (2.5–5 cm)

Spacing:
4–6 in. (10–15 cm)

Days to germination:
7–14 days

Days to maturity:
60–75

As with most noodle beans, 'Chinese Red Noodle' is best eaten before the seeds fully fill out.

The first thing you will notice when you bite into a fresh 'Dragon's Tongue' bean is the cool juiciness that practically dribbles down your chin. This pale yellow wax bean—beautiful to behold—is etched with purple streaks that disappear when cooked. Unlike other wax beans that have far more texture than flavor, this is no anemic yellow bean.

Do not be fooled by the compact size of the plants, which will continue to produce throughout the summer. These versatile beans can be used as snap, shelling, or dried beans, but they are hard to resist while fresh and in all their colorful glory. The wide pods can cover the plant, and they stay stringless as they mature, changing from a pale green to an icy, almost translucent yellow. 'Dragon's Tongue' was part of a concerted effort in the Netherlands to develop a sweeter wax bean. You can also find this Dutch heirloom listed as 'Dragon Langerie' and 'Merveille de Piemonte'.

BEAUTIFUL

Bean 'Dragon's Tongue'

'Dragon Langerie', 'Merveille de Piemonte'

Phaseolus vulgaris

Flavor Despite its pale coloring, 'Dragon's Tongue' has a bright, green bean flavor with a subtle sweetness in the juice that is lively and refreshing. It makes a tasty fresh, raw snack and can be used in any recipe calling for green beans. Do not overcook them or they will lose their bright flavor. They make great pickles, too.

Growing notes Direct seed in the garden after all danger of frost has passed. You can succession plant every 2–4 weeks to stagger and prolong the harvest. Gardeners in frost-free climates can continue planting into the fall.

The short, bushy plants should not require support, but give them room to spread out.

They are light feeders and do not need supplemental fertilizer if the soil includes plenty of organic matter. These are rust-resistant and not bothered by many bean pests.

How to harvest Harvest when the beans are about 6 in. (15 cm) long and feel full and firm to the touch. Snap beans are ready when they have developed purple streaks. As the streaks turn more reddish, you can harvest the beans for shelling.

Tip 'Dragon's Tongue' is a good choice for growing in containers. They are small enough to produce in a confined space, yet pretty enough to be part of an ornamental mix.

Others to try 'Brittle Wax' is sweet with an excellent crunch. 'Cherokee Wax' won an All-America Selections award in 1948. 'Pencil Pod Golden Wax' is a vigorous grower, with tender, long, and stringless beans that are great for freezing or canning.

Exposure:
Full sun

Ideal soil temperature:
75–85°F (24–30°C)

Planting depth:
1–1½ in. (2.5–4 cm)

Spacing:
2–3 in. (5–7.5 cm)

Days to germination:
7–10 days

Days to maturity:
55–60 days

'Dragon's Tongue' are creamy, purple-streaked beans that produce heavily for weeks on end.

LEGUMES

Slender and crisp as a matchstick, 'Fin de Bagnol' is the classic French filet bean. The *Fin* in the name means fine or thin. 'Fin de Bagnol' was also known as the shoestring bean, because it is long and lanky and tapers at the end, like a long string. It has been a favorite in American gardens since the 1800s.

Filet beans, also known as *haricot verts* (green beans) are harvested when young and thin. It is not easy being skinny and green; as they put energy into the seeds, bean pods tend to fill out and plump up. To catch them at their finest, you need to harvest them every couple of days. They mature quickly and, although they are small plants, they produce prodigiously. These beans do not store well, so enjoy them soon after picking.

Exposure:
Full sun

Ideal soil temperature:
75–80°F (24–27°C)

Planting depth:
1–1 1/2 in. (2.5–4 cm)

Spacing:
4–6 in. (10–15 cm)

Days to germination:
5–10 days

Days to maturity:
50–55 days

'Fin de Bagnol' pods are slender and can be difficult to find among the leaves, even though they grow in large masses.

Bean 'Fin de Bagnol'

French filet bean, shoestring bean, haricot verts
Phaseolus vulgaris

Flavor 'Fin de Bagnol' are tender and crisp with a complex flavor—the expected green bean liveliness, tempered with a nutty sweetness. The brittle tenderness makes this selection distinctive. Harvest well before the seeds have developed inside and you can taste the dynamic, sharp bean flavor of the velvety pods. Blanch them quickly to retain their crispness and flavor.

Growing notes Direct seed in the garden after all danger of frost has passed. 'Fin de Bagnol' can withstand cooler temperatures, but the seeds require dry soil to avoid rot. Succession plant every 2–3 weeks. Gardeners in frost-free climates can sow seeds in early fall.

These short, bush type plants should be able to support themselves. Leave room between plants to allow air circulation and to help you locate the skinny beans among the foliage. The plants are shallow rooted and need regular watering to taste their finest. Plants grow quickly and should not require supplemental fertilizer if the soil is rich in organic matter. Mulching around the plants will keep their roots cool and help conserve water.

Mature plants are not often bothered by pests, but the young seedlings are easy prey for insects and other creatures. Insects can be foiled by placing a row cover over the seedlings until the plants start to bloom. You can use a barrier to keep out animals such as rabbits and groundhogs. Wire fencing at least 3–4 ft. (0.9–1.2 m) tall with a 1 in. (2.5 cm) mesh will do the trick, especially if you bury an additional 6–12 in. (15–30.5 cm) of the material around the perimeter of the garden.

How to harvest Harvest while the beans are shoestring thin, about 4–6 in. (10–15 cm) long, and bright green. Because the pods blend in with the stems and leaves, you will need to do some hunting to find them all, but be persistent, because these beans get tough and stringy if they remain on the plant too long.

Tip The warmer the weather, the more frequently you need to harvest.

Others to try 'Beurre de Roc-quencourt' is a buttery yellow filet bean. 'Finaud' stays tender longer, so you can harvest less often. 'Triomphe de Farcy' is marbled with purple and matures slightly later than 'Fin de Bagnol'.

The wonder of 'Kentucky Wonder' is that these beans can remain on the vine for a long time without getting tough. Pods of 8–10 in. (20–25 cm) stay crisp, yet tender and are relatively stringless. Once sold as Texas pole beans, 'Kentucky Wonder' is popular particularly in the southern United States, where it is also known as 'Southern Prolific' and 'Old Homestead'. Many recipes still use one of these names in the ingredients list.

Pole beans are great choices for home gardeners because they produce for weeks at a time, offering a great yield in a small space. 'Kentucky Wonder' is not only a productive bean, but one of the earliest to mature.

CLASSIC

Bean 'Kentucky Wonder'

Texas pole bean
Phaseolus vulgaris

Flavor 'Kentucky Wonder' is often described as having a distinctive flavor—a clean, almost acidic edge to the nutty bean taste we expect from green beans. It is a wonderful cooked green bean that freezes well. If you neglect to harvest for a week or two, you can still harvest the plump, mature pods as shelling beans. Some cooks use the shelled beans to make baked beans.

Growing notes Direct seed in the garden, after all danger of frost has passed. Bean seeds will rot in wet soil, so wait for the ground to dry out and feel crumbly in your hand. Gardeners in frost-free climates can plant a second crop in midsummer to late summer.

Pole beans require support, which should be in place when you plant, rather than waiting until the plants have established roots, which can be damaged if disturbed. 'Kentucky Wonder' beans are light feeders and should not require supplemental fertilizer if the soil includes plenty of organic matter. Keep the plants well-watered.

How to harvest Harvest when the beans are about 6–8 in. (15–20 cm) long and feel full and firm. Beans are easily snapped off the vine, but hold the vine end as you snap to avoid taking part of the vine with you. Har-

Exposure:
Full sun

Ideal soil temperature:
75–80°F (24–27°C)

Planting depth:
1–1¹/₂ in. (2.5–4 cm)

Spacing:
2–3 in. (5–7.5 cm)

Days to germination:
8–16 days

Days to maturity:
58–72 days

'Kentucky Wonder' is a classic that is not only one of the most productive beans, but also one of the earliest.

vest every day or two and the vines will continue to produce.

Tip Although 'Kentucky Wonder' remains tender as it matures to its full 8–10 in. (20–25 cm), the beans will be more fully stringless if they are harvested young, at about 4 in. (10 cm) long.

Others to try 'Black Valentine' is a bush variety that holds its flavor after harvesting. 'Purple Podded Pole' is a meaty bean with high yields of reddish-purple pods. 'Sultan's Green Cresent' curls as it grows but stays stringless and tender.

'Lazy Housewife' has more to recommend it than just its intriguing name. It has the distinction of being the first completely stringless bean with a lovely crisp snap. The pods have the obliging habit of growing in clusters, so you can harvest more beans in less time, without having to hunt through a lot of leaves. 'Lazy Housewife' beans are well-named: they are less work for you and offer all the rewards of a remarkable green bean.

Bean 'Lazy Housewife'

Stringless green bean
Phaseolus vulgaris

Flavor These beans are delectable when eaten raw or just warmed through; they are so crisp that they almost shatter in your mouth. Their bright, slightly astringent, floral flavor is enhanced by the strong bean scent released when you bite into one. That multisensory experience richens the liveliness of fresh beans. They are best used when cooked only slightly, such as in bean salads.

If you take the effort of drying and cooking the seeds, you will be rewarded with beans of an entirely different character. When dried, they have a dark, earthy, lentil-like flavor.

Growing notes Direct seed in the garden, after all danger of frost has passed and the soil has dried and warmed to about 70°F (21°C). This is a long-season bean, and succession planting in cooler climates may prove futile. However, gardeners in frost-free climates can plant a second crop in midsummer to late summer.

Vines can climb to 8 ft. (2.4 m) and become heavy with clusters of bean pods. Install a support or trellis when you plant; you might find it difficult to set it up later, because pole beans cling to whatever is nearest them.

Although these beans are light feeders, plant them in rich soil. Their long season warrants a side dressing of compost or manure in the middle of the growing season. The plant is fairly disease resistant, growing and producing well in most weather conditions. Water the plants regularly, at least 1 in. (2.5 cm) per week or more during dry spells.

How to harvest For snap beans, harvest when the bean pods are about 6–8 in. (15–20 cm) long and feel full and firm. Keep harvesting every day or two and the vines will keep producing. Shelling beans should remain on the plants until the pods are completely dry.

Tip 'Lazy Housewife' is a slow starter, so plant an early variety or a row of bush beans such as 'Bountiful' or 'Contender' to harvest before the slower beans are ready.

Others to try 'Cherokee Trail of Tears' is a historic pole bean with rich black seeds and is good fresh or dried. 'Rattlesnake' is a productive, early snap bean for fresh eating. 'Supermarconi' is a flavorful bean that is not affected by cool, wet weather.

Exposure:
Full sun

Ideal soil temperature:
75–85°F (24–30°C)

Planting depth:
1–1½ in. (2.5–4 cm)

Spacing:
2–3 in. (5–7.5 cm)

Days to germination:
8–16 days

Days to maturity:
75–90 days

The clustering habit of 'Lazy Housewife' makes it easy to grab a handful of delicious beans.

Eating 'Romano' beans is like treating yourself to some warm and comforting pleasure. The flattened pods are so tender and decadently rich, you might feel guilty eating them. Indulge yourself.

'Romano' is a multipurpose bean. Its delicate flavor holds up wonderfully well when canned or frozen. Young beans start off looking like average string beans. As they mature, however, they get wider and flatter, except for the protruding seeds. If you leave them on the vine to ripen, they also make a great shelling bean. These beans are too perishable to be found in the grocery produce aisle, so grow them yourself to enjoy their freshness.

Americans have not always been fond of 'Romano' beans. Apparently seed companies in the early 20th century did not want to advertise their unusual shape, so no images of the beans were used on seed packages. Imagine the surprise of the unsuspecting gardeners who planted them.

Exposure:
Full sun

Ideal soil temperature:
75–80°F (24–27°C)

Planting depth:
1–1 1/2 in. (2.5–4 cm)

Spacing:
4–6 in. (10–15 cm)

Days to germination:
7–14 days

Days to maturity:
70–80 days

'Romano' beans are fat and flattened, even when the beans are mature.

Bean 'Romano'

Italian green bean
Phaseolus vulgaris

Flavor Tender and meaty, 'Romano' beans have a rich flavor unto themselves, with little of the sharp, astringent taste associated with green beans. Their delicate creaminess makes them very satisfying to eat, like mashed potatoes, and helps them blend well in other dishes, such as soups, stews, and pasta.

Growing notes Direct seed in spring, after the soil has warmed to at least 50°F (10°C). Bush 'Romano' can be succession planted every 2–3 weeks, into midsummer. The pole variety can be planted once or twice in the spring and early summer. Gardeners in frost-free climates can make successive plantings in midsummer to late summer.

Pole 'Romano' requires some type of support, which you should install when you plant. The pole variety will out-produce bush 'Romano', but the bush variety will start producing earlier.

This is an easy grower that needs no supplemental fertilizer if planted in soil rich with organic matter. These beans are generally one of the last to be eaten by bean beetles, and they are resistant to mosaic virus.

How to harvest For eating fresh, harvest the pods when young, at about 3–5 in. (7.5–12.5 cm) long and before the pods appear lumpy. You can also let the pods fill out and use them as fresh shelling beans, or leave them on the vine to dry and shell as dried beans.

Others to try 'Gold of Bacau', from Bacau, Romania, are sweet, flat, and tender, but crunchy. 'Musica' pole beans are a meaty, prolific Spanish variety. The violet pods of 'Romano Purpiat' turn green when cooked; they share the same tender sweetness of 'Romano' and are good for gardens in cooler regions.

One look at the bold, burgundy and white 'Christmas' lima bean tells you this is no ordinary bean. These large lima beans pack a lot of flavor in their dense package. Among its many aliases, 'Christmas' is also called the chestnut lima bean because of its sweet, nutty flavor; calico lima, for obvious reasons; and giant Florida pole because it is especially popular with Florida gardeners. It can handle heat and humidity well and was a favorite in the southwestern United States as far back as the 1840s.

Lima beans were so named on export to Europe and the Americas from Lima, Peru; they have been grown for centuries in the Andes and throughout Mesoamerica. 'Christmas' is a slow-growing pole bean that needs a long, warm growing season, and this prevents it from becoming a viable commercial bean. You must patiently wait until the end of the season to harvest these beans and experience the rich, satisfying flavor that can be the centerpiece of a meal. 'Christmas' lima beans deliver with flavor, substance, and beauty.

Lima bean 'Christmas'

Chestnut lima, calico lima, butter bean

Phaseolus lunatus

Flavor 'Christmas' lima beans taste sweet and nutty, with a rich, starchy flavor quite unlike more common green lima varieties. The fluffy potatolike texture combines with the nutty flavor, allowing it to shine on its own with the simplest of preparations, and it mingles well with hearty seasonings, such as curries and garlic.

Growing notes Direct seed in the garden about 2–3 weeks after the last frost date. They need a long growing season, and gardeners in cooler areas can start their plants indoors. 'Christmas' lima beans stand up well in heat. They can also be seeded in the fall in frost-free climates.

The vigorous vines can easily reach heights of 8–10 ft. (2.4–3 m); add a strong trellis or teepee at planting time, and plant five or six seeds to each support. Because of their long growing season, a dose of compost or manure in midsummer will give the beans an energy boost. They also need plenty of water. Flowers will drop off the plants in hot, dry conditions.

Watch for the usual bean pests, such as aphids, bean beetles, and flea beetles, as well as mosaic virus. Young lima beans are especially susceptible to pests.

How to harvest Harvest these limas to eat fresh, when the beans have plumped and the pods feel full. The pods should be a bright green and the ends will feel spongy.

Tip Pot liquor is the liquid in which lima beans, peas, and greens are cooked. When you cook 'Christmas' lima beans, instead of draining and disposing of the pot liquor, you can use it to flavor the rest of the dish.

Others to try 'Alabama Black-

Exposure:
Full sun

Ideal soil temperature:
75–85°F (24–30°C)

Planting depth:
1–1½ in.
(2.5–4 cm)

Spacing:
4–6 in. (10–15 cm)

Days to germination:
7–10 days

Days to maturity:
75–100 days

'Christmas' is a wonderfully satisfying bean to eat, but it requires a long growing season.

eyed' has a rich, nutty flavor. 'Dixie Speckled Butterpea' is tiny, sweet, and tender, but difficult to shell. 'Jackson Wonder' is a bush variety that can be harvested young and is good for areas with shorter growing seasons.

The beautiful asparagus pea, also known as the winged pea or winged lotus, is not a pea (*Pisum*), a water lotus (*Nelumbo*), or an asparagus. It is an intriguing little plant that thrives in poor soil and produces delectable green pods with flaps or wings on four sides, much like the 'Hunan' winged bean (*Psophocarpus tetragonolobus*). The foliage and bright red flowers are somewhat pealike and ornamental enough for the asparagus pea to be grown in border plantings.

Asparagus pea is thought to be a native of northwest Africa, but it came to prominence in the Mediterranean region, where it was enjoyed as a seasonal treat. We know it was sold in Philadelphia in the early 1800s. Back then, it was called the winged pea. Food writer William Woys Weaver said the Philadelphia gentry used it as a garnish on fricasseed frog legs, but we do not need frog legs to enjoy asparagus peas.

LEGUMES

UNUSUAL

Pea, asparagus
Winged pea, winged lotus, winged bird's foot trefoil
Lotus tetragonolobus

Flavor The asparagus pea tastes slightly of asparagus, with a peppery sweetness that makes the flavor distinctive—all the more so when they are cooked and the aroma suffuses the air. Gentle cooking methods, such as steaming, are best for retaining their texture and quality. They can be a bit overwhelming on their own, but they liven up other vegetable dishes when added.

Growing notes Start seeds indoors, about three weeks before the last expected frost date. You can also direct seed in the garden after all danger of frost has passed if you keep the soil constantly moist. Asparagus pea is grown as an annual, and gardeners in frost-free areas can start them at any time. Succession plant monthly to keep the harvest going.

Germination is poor, so you will need to start more seeds than you need. Although the plants are tough enough to grow in poor soils, they will be happier in a moderately rich soil with plenty of water, at least until established. Once growing, the plants take care of themselves. Asparagus peas can sprawl out on the ground or tumble over a container.

Asparagus peas do not cross-pollinate with other plants; to save seed, simply let the pods dry on the plants, remove the seed, and store in an airtight container for next season. (And remember to share them with friends.)

How to harvest Harvest the pods while young, at less than 1 in. (2.5 cm) in length. They can get tough as they mature, and they mature quickly.

Tip You can plant asparagus peas in containers, and they will drape over the edges of the pot

Exposure:
Full sun to partial shade

Ideal soil temperature:
65–75°F (18–24°C)

Planting depth:
1/4–1/2 in. (0.6–1 cm)

Days to germination:
7–14 days

Spacing:
12–18 in. (30.5–45.5 cm)

Days to maturity:
60–75 days

The bright red flowers of the asparagus pea are as attractive as the unusual winged pods.

quite prettily, with dark red blossoms and winged seed pods. A 12 in. (30.5 cm) container is adequate for a single plant, but pods come in erratically, so you might want to include more plants in the mix.

Others to try 'Hunan' winged bean (*Psophocarpus tetragonolobus*) is similar in appearance with more of a bean flavor. These tall vines require at least 2 months of warm night temperatures to flower and produce. Ostrich fern (*Matteuccia struthiopteris*) fiddleheads are small, spiraled shoots that appear briefly in early spring and offer a crunchy treat with a flavor reminiscent of asparagus.

When the sun hits these peas, the translucent, lemon-yellow pods show a hint of the tiny peas inside. This gorgeous plant has vivid lilac flowers and golden stems cupped by leaves ringed in a rosy magenta. It is also a prodigious producer of the peppiest, most scrumptious and crunchy golden pea pods.

Perhaps the coloring should be a hint that 'Golden Sweet' hails from India, land of colorful spices and dyes, where it has been traced back to the 1600s. A fringe benefit of all that color is that the plentiful pods are easy to find among the leaves.

BEAUTIFUL

Pea 'Golden Sweet'
Snow pea
Pisum sativum

Flavor 'Golden Sweet' is one of the crispest and most succulent snow peas, with a flavor more full-bodied than that of the average pea. They can be enjoyed fresh and raw or gently cooked. Along with its sugary juiciness, it offers all the flavorful character of a mature pea, with a slightly spicy aftertaste.

Growing notes Peas like cool weather, so sow seeds early in the spring, after the ground has warmed to about 45°F (7°C) and the soil is dry enough so that when you squeeze it into a ball it falls apart with a tap of your finger. A second planting can occur in midsummer to late summer. Gardeners with warm springs in Zones 7 and warmer will get best results planting peas in the fall. 'Golden Sweet' will grow tall before it starts to flower.

The vines can reach 6 ft. (2 m) tall and are full, top to bottom, with pods. Attach the vines to a tall, study trellis or fence to keep them off the ground and from

collapsing. Allow for plenty of air circulation or the pods will be prone to disease and rot. Rotting or fungus problems can occur during unusually wet weather; good air circulation is key, and trellising and the return of sunshine will help control these problems. Aphids can also be a problem, especially if the vines get thickly entwined or collapse.

Pea roots fix nitrogen and plants do not require supplemental fertilizer. Cut the spent vines at the soil line to allow the roots to remain and continue to feed the soil.

How to harvest 'Golden Sweet' is tolerant of some heat and should produce well into summer. It is best when harvested before the peas start to plump up. If you harvest them young, they are virtually stringless, but even if you miss a few, older pods will be crisp with good flavor.

Tip 'Golden Sweet' peas can be left on the vine to mature; shell, dry, and store the peas to use in soups throughout the winter.

Others to try 'Mammoth Melting Sugar' has shorter vines with a large yield and truly sweet pea pods. 'Oregon Sugar Pod' is

Exposure:
Full sun

Ideal soil temperature:
65–75°F (18–24°C)

Planting depth:
1–1½ in. (2.5–4 cm)

Days to germination:
7–14 days

Spacing:
2 in. (5 cm)

Days to maturity:
65–75 days

LEGUMES

'Golden Sweet' stays crisp, even as the tiny peas begin to fill out.

a popular bushy plant with sweet peas that stay tender even as they plump up, but it is not heat tolerant. 'Sugar Snap' is very sweet and quick to yield.

Pleasantly sweet, plump, and prolific, 'Tall Telephone' peas are so tantalizing, you may eat them before you reach the house. Although they are considered shelling peas, they make a wonderful cool snack to be enjoyed straight off the vine as you work in the garden.

In 1881, five years after Alexander Graham Bell spoke the immortal words, "Mr. Watson, come here. I want to see you," 'Tall Telephone' peas were introduced, named in honor of the invention that was sweeping the nation. This tall, old-fashioned variety is a sight to see when laden with masses of plump pods. 'Tall Telephone', or 'Alderman' as it is known in England, reaches 6–8 ft. (2–2.5 m) tall and drips with hundreds of pods.

Exposure:
Full sun

Ideal soil temperature:
65–75°F (18–24°C)

Planting depth:
1–1 1/2 in. (2.5–4 cm)

Days to germination:
7–14 days

Spacing:
2 in. (5 cm)

Days to maturity:
65–75 days

'Tall Telephone' peas are seriously tall vines; do not skimp on the trellis.

Pea 'Tall Telephone'

Pole pea, 'Alderman'
Pisum sativum

Flavor This pea offers a sweet, succulent, juicy burst of flavor when eaten fresh and holds its flavor and texture well when frozen. The peas themselves are the wrinkled type, which means less starch and more pleasant sweetness to soften their savory flavor.

Growing notes 'Tall Telephone' produces best after a long, cool season, so start seeds as soon as the ground has warmed to about 45°F (7°C) and the soil is dry enough that it does not stick to your tools. Peas give out quickly in heat, so keep them cool with plenty of water. A fall crop can be seeded in midsummer to late summer. Gardeners with warm springs in Zones 7 and warmer will get best results planting peas in the fall.

As you plant the seeds, erect a sturdy trellis to keep the vines off the ground and allow for plenty of air circulation or the pods will be prone to disease and rot.

This pea produces a heavy crop in a short time and is a fairly easy grower, but it will stop producing when the temperatures warm and stay above 80°F (27°C). In damp seasons, rotting or fungus diseases can be problematic, but these can be controlled by good trellising and the return of sunshine.

Pea roots fix nitrogen and do not require additional fertilizer. Cut the spent vines at the soil line and allow the roots to remain and feed the soil.

How to harvest Harvest peas as soon as the pods feel full. Gently squeeze the pods to test for plumpness. The younger they are harvested, the sweeter they will be. Harvest regularly to encourage the vines to produce more peas.

Tip 'Tall Telephone' sets most of it pods high on the vines and can get top heavy and droop when loaded with pods. Harvest often and secure the vines to a strong trellis or other support.

Others to try 'British Wonder' vines are shorter and produce prolifically. 'Carlin' (or 'Carling') grows on tall vines and is usually dried for later use. 'Prince Albert' produces tall vines with yellow-tinged peas and is popular for soups.

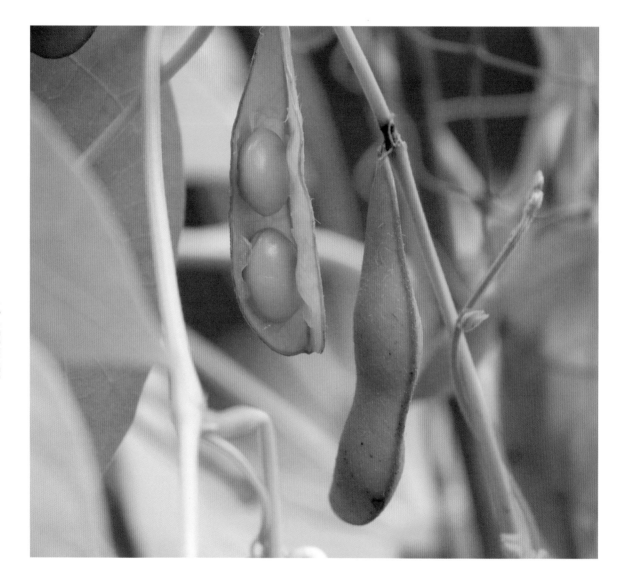

'Envy' soybeans make a nutty, buttery snack, with smooth, emerald-colored beans that can be popped right into your mouth. They are a short season variety, maturing in less than 3 months, which makes them perfect for home gardeners.

This is not an old heirloom. It is the work of plant breeder Elwyn M. Meader, who experimented with delicious garden favorites to make them more reliable to grow. During his tenure at the University of New Hampshire and well after his retirement in 1966, Meader developed many of the classic vegetables we still grow today.

You have probably seen soybeans sold as Japanese *edamame* (branched bean). I think the Chinese name *mao dau*, hairy bean, is even more descriptive, because the pods are quite fuzzy. The vivid green plants reach only about 24 in. (61 cm) tall and can be grown in containers. Three or four plants can easily be packed into a 12 in. (30.5 cm) pot.

Soybean 'Envy'
Soya bean
Glycine max

Exposure: Full sun

Ideal soil temperature: 70–80°F (21–27°C)

Planting depth: 1 in. (2.5 cm)

Spacing: 3–4 in. (7.5–10 cm)

Days to germination: 7–14 days

Days to maturity: 75–85 days

'Envy' grows quickly enough to produce beans in just about any climate.

Flavor 'Envy' tastes like a sweet, nutty, lima bean, with a firm, smooth texture. The firmness is retained in cooking because of its high protein, low starch content. These soybeans are a nice size for snacking. Boil them in salted water for 7–10 minutes and eat them right out of the pod. They are also great pureed and mixed into sauces and pesto.

Growing notes Direct sow seed about 2–3 weeks after the last frost date. Gardeners in cool climates can get a head start by starting seeds indoors a month earlier. You can also succession sow every 2–3 weeks throughout the summer to prolong the harvest.

Soybeans grow much like regular green beans. They prefer warm soil, at least 70°F (21°C), that is moist and rich in organic matter. Soybeans can withstand drought and fix their own nitrogen. Mulching will keep the roots cool and help conserve water.

'Envy' should not need staking, although the plants can get lanky. Plant them close together so each plant can support another.

The usual bean diseases and pests, particularly beetles, may affect soybeans. Preventing stress from too much or too little water will make them less susceptible to problems.

How to harvest Harvest the pods when they feel plump and are still a bright green. Each pod will hold two or three beans.

Tip Fresh beans can be difficult to shell. Steaming or blanching them will help them pop open, like peas.

Others to try 'Agate' is a tasty New Mexico heirloom that is also early to mature. 'Fiskeby' is another early variety, bred in Sweden for cooler climates. 'Shirofumi' is a delicious, heavy yielding, vining soybean.

113

LEGUMES

MELONS

Growing melons is like growing dessert. Fresh melons are aromatic, superbly sweet, and juicy. The heavenly aroma of a vine-picked melon will fill your senses. Some slip off the vine easily in your hand, and some signal it is time to eat by their subtly changing color.

Melons kick into high gear when the summer heats up. All they ask is plenty of sunshine, water, and room to spread. Give them what they need, and your reward will be ambrosia.

‘Boule d'Or' is a gentle, sweet melon with a striking color.

It is easy to see where 'Boule d'Or' gets its name: it truly is a gold ball. This old French heirloom has been marketed since the mid-1800s. Its luscious, pale green honeydew flesh drips with a sweetness modern honeydews cannot match. The bold, gold skin has just the faintest netting.

SWEET

Exposure:
Full sun

Ideal soil temperature:
75-85°F (24-30°C)

Planting depth:
1/2-1 in. (1-2.5 cm)

Days to germination:
5-10 days

Spacing:
Plant 2 or 3 seeds in hills 4-6 ft. (1.2-2 m) apart and thin to best plant per hill

Days to maturity:
95-110 days

Melon 'Boule d'Or'

Honeydew 'Golden Perfection'
Cucumis melo

Flavor The color alone will draw you to this melon, but slice into one and you will be treated to the combination of aroma and juicy goodness. The fragrance tells you this is no ordinary melon: its scent is somewhere between confection and floral and the flavor is just as complex. It positively envelopes your senses. To get the full impact, be sure to let it ripen fully on the vine.

Growing notes Direct sow seeds in spring; wait until the soil temperature warms to 60-65°F (15-18°C) before sowing. Exposure to colder temperatures can mean no melons all season. You can also start seed indoors, 3-4 weeks before transplant date; start in peat or paper pots, because melons do not like their roots disturbed.

'Boule d'Or' needs a long, hot season, but gardeners in Zones 8 and warmer can squeeze in a crop in midsummer to late summer to harvest in late fall.

Because it is a full-season plant, it needs a rich soil that drains well and lots of sunshine. Floating row covers can be used early in the season to keep the plants warm and the bugs out. Remove the covers when flowers start forming so bees can pollinate the plants.

Water transplants well to get them established. Provide regular water to the plants throughout the growing season, until the melons start ripening. Withholding some water as they ripen helps concentrate the sugars.

How to harvest Honeydews are tricky to harvest. They do not detach from the vine to tell you they are ready. 'Boule d'Or' will be a glowing yellow and the netting will be barely noticeable without touching it. You should be able to detect a whiff of fragrance when it has peaked.

Melons do not continue ripening after they are picked, so for the best flavor, harvest at their peak. Honeydews will keep longer than cantaloupes. Check for freshness by pushing the stem end to see if it has some give to it. You might notice some cracking of the stem near where it joins the melon.

Others to try 'Charentais' is a luscious French cantaloupe, but a finicky grower. 'Green Nutmeg' is an aromatic, spicy, green-fleshed cantaloupe that delivers high yields. Orange flesh honeydew is a deceptive color for a honeydew but is exceptionally sweet.

'Minnesota Midget' is a backyard gardener's dream. These super-sweet, softball-sized muskmelons grow on unusual 3–4 ft. (0.9–1.2 m) vines. You can even grow 'Minnesota Midget' in a container on your patio, because it is that well behaved. Each plant will produce six to eight melons, each perfect as a treat for one or two people.

This short-season grower is a good choice for gardens in cooler climates, but it will thrive in almost any climate. The Minnesota moniker simply refers to where it was bred and introduced.

Exposure:
Full sun

Ideal soil temperature:
75-85°F (24-30°C)

Planting depth:
1/2-1 in. (1-2.5 cm)

Days to germination:
5-10 days

Spacing:
Plant two or three seeds in hills 2-4 ft. (0.6-1.2 m) apart; thin to best plant per hill

Days to maturity:
60-70 days

AROMATIC

Melon 'Minnesota Midget'

Cantaloupe, muskmelon
Cucumis melo

Flavor The thick, orange walls of this melon pack a lot of flavor into a small package. A high sugar content makes it sweet enough for dessert, and the firm, supple texture truly melts in your mouth. These tiny gems have thin rinds, so no dewy flesh is wasted.

Growing notes Do not plant seeds outdoors until the soil temperature reaches 65-75°F (18-24°C). You can also start seed indoors, 3-4 weeks before transplant date, but melons do not like their roots disturbed, so start in peat or paper pots that can be transplanted directly

A ripe 'Minnesota Midget' melon absolutely glows, and the fragrance tells you it is time to harvest.

in the ground. In Zones 8 and warmer, you can make a second planting in midsummer for a fall harvest.

Water transplants well to get them established. Provide regular water to the plants throughout the growing season, until the melons start ripening. Withholding some water as they ripen helps concentrate the sugars.

'Minnesota Midget' is a fast grower, so do not be tempted to start the growing season too early. Like any melon variety, it needs rich, well-draining soil and lots of sunshine. It is a good choice for containers, with

vines growing to about 4 ft. (1.2 m) long. A container will help keep the roots warm and lift the fruits off the ground, away from insects and four-legged pests. Remember that soil in containers will dry out quickly, so extra water will be required.

This melon is resistant to *Fusarium* wilt, a true bane of melon growers.

How to harvest Melons will emit a sweet scent as they near maturity. Muskmelons are harvested when at full slip, which means they separate from the vine with the slightest effort. You will probably notice some cracking of the stem near where it joins the melon. Because of this, you should provide support for fruits grown on a trellis. Cradle each fruit on the vine with something that has a bit of give, such as old stockings or netting.

Others to try 'Eden's Gem' produces softball-sized green melons. This Colorado selection is extremely sweet and juicy. 'Green Climbing' (syn. 'Vert Grimpant') is a fragrant French heirloom with small green fruits. It prefers a hot, dry climate. 'Petit Gris de Rennes' is an old, sweetly perfumed French melon.

Fragrant, sweet, and crunchy, 'Blacktail Mountain' is everything a great watermelon should be. On the outside, it looks like a glossy, green bowling ball. Slice into one and you can hear the crispness of the hearty scarlet flesh.

This is a new, or created, heirloom. As a teenage gardener, Glenn Drowns, now of the Sand Hill Preservation Center (an heirloom seed preservation organization), was determined to get a watermelon to ripen during the short Idaho growing season. He spent several years cross-breeding before achieving the wonderful accomplishment of 'Blacktail Mountain'. What a delight it must have been not only to create an early maturing watermelon, but one with such exceptional flavor.

This melon is ideal for cooler climates and short growing seasons, but it is also unfazed by heat and humidity, which is more than most of us can claim.

Watermelon 'Blacktail Mountain'
Citrullus lanatus

Flavor The flesh is dense but juicy, with a bright sweetness that mingles with the fruity aroma and conjures up images of back porches and summer afternoons. One chilly slice and you will forget all about the hazy heat and humidity.

Growing notes Wait until the weather has warmed to about 70°F (21°C) before direct sowing. Sowing in cooler temperatures can prevent the plant from setting fruit the entire season. Seedlings can be started indoors about 3 weeks before setting out. Use peat or paper pots so the roots are not disturbed when transplanting. Heat-loving watermelons are not usually grown in the fall, but gardeners in Zones 8 and warmer can plant in midsummer for a later harvest.

Melons, like squash, are often planted in hills, a small mound in which 3 to 5 seeds are buried. Once the seeds sprout, thin the plants to the two strongest. Planting multiple plants in close proximity improves pollination.

Exposure:
Full sun

Ideal soil temperature:
75-85°F (24-30°C)

Planting depth:
1-1½ in. (2.5-4 cm)

Spacing:
1 ft. (0.3 m) apart, or 3-5 seeds per hill, 3 ft. (0.9 m) apart

Days to germination:
5-14 days

Days to maturity:
70-75 days

'Blacktail Mountain' looks like a glossy green bowling ball but tastes like a dream.

Watermelons can be slow starters: they will sprout, put out a few leaves, and then mope for awhile, so be patient. As soon as the weather heats up, they will get moving again.

Keep the plants well watered and give them plenty of room to sprawl. The more sun they get, the better. Ease up on watering as the fruits near maturity to help concentrate and intensify the sugars.

How to harvest Watermelons will not continue to ripen after harvesting, and it can be difficult to judge the best moment to pick them. Watch for a few signs:

The tendrils near the stem turn brown; a pale yellowish spot appears where the melon touches the ground (the spot is white and turns yellow as the melon ripens); the rind color looks dull, and when you press your nail into the rind, it does not easily sink in.

Others to try 'Crimson Sweet' is crisp, sweet, and reliable in frost-free climates. 'Moon & Stars' is a watermelon legend for its beautiful markings and sweet, pink flesh. 'Sugar Baby' produces small, sweet melons and lives up to its name.

OKRA

Okra is a stunning plant that can do double duty as an ornamental and an edible plant. It is related to the ornamental hibiscus, which is obvious by its tropical-looking flowers. The young, tender pods are closely identified with Creole, Cajun, and Southern cooking; when cooked, their gooey texture works well in gumbos and stews.

     Okra was discovered in the 12th century BC in Africa. The ancient Egyptians were known to have enjoyed okra, and throughout North Africa and the Middle East, not only were the pods used as vegetables, but the roasted seeds were used as a coffee substitute. Okra is still an amazingly versatile vegetable when eaten fresh or cooked, pickled or fried, alone or perfectly paired with corn, onions, tomatoes, or peppers.

Many plants are beautiful and delicious, but I challenge any to top 'Burgundy' okra. It touts beautiful flowers and flushed foliage, and, unlike most okra varieties, it has no spines. The rich, red pods retain their tender crunch as they elongate. 'Burgundy' has a nutty sweetness that adds a unique character to common dishes. All that would be good reason to grow this okra, but it also happens to out-produce most every other okra variety, and okra plants are not slackers in general.

'Burgundy' okra is a newcomer to the heirloom fold. It was developed at South Carolina's Clemson University in the early 1980s and was a 1988 All-America Selections winner. Although it does not meet the somewhat arbitrary guideline of being at least 50 years old, it is open pollinated, developed with the help of some heirloom parents, and it has been embraced by heirloom growers. It is also a great example of how the category of heirlooms is dynamic and ever-expanding. Old, standard varieties are always being improved, sometimes by accident and sometimes by intent. If you are lucky, the result can be an instant classic.

LONG SEASON

Exposure:
Full sun

Ideal soil temperature:
75–85°F (24–30°C)

Planting depth:
1/2–1 in. (1–2.5 cm)

Days to germination:
7–14 days

Spacing:
10–12 in.
(25–30.5 cm)

Days to maturity:
70–80 days

Okra 'Burgundy'
Abelmoschus esculentus

Flavor 'Burgundy' has a dense, earthy flavor and a creamy crunch. It is less fibrous than the green varieties and remains tender even if you miss a day's harvest and it grows a bit large. Although okra is often used to soak up the other flavors in a dish, this okra's distinct nuttiness complements other foods. It does lose some of its beautiful color with cooking, however.

Growing notes Gardeners in Zones 6 and cooler can direct seed in the garden, 2–4 weeks after the last frost date. You can also start seed indoors, 2 or 3 weeks before transplanting. Even in frost-free climates, okra is usually planted in the spring.

Because it loves the heat, it will keep producing throughout the summer and into fall.

Okra seeds have a hard shell, and soaking them overnight will help speed germination. Do not put seeds and plants outdoors until the temperature is reliably above freezing, or you will lose both. Okra does not transplant well, so if you are starting seeds indoors, use a peat or paper pot to prevent transplant shock.

Okra is a heavy producer and likes a fertile, well-drained soil. A little extra fertilizer when the plants first start producing will give them an added boost. Otherwise, water during dry spells and keep harvesting. Some pests will munch on okra leaves, but few bother with the pods.

How to harvest Although pods stay tender longer than other varieties of okra, they are still at their best when harvested

at 2–4 in. (5–10 cm) long. Pods grow quickly, so check them often. Cut off the pods rather than pulling them to avoid damaging the plants.

Tip 'Burgundy' okra is not quite as "sticky" as other okras, and to minimize this even more, use it freshly picked and cook it quickly and lightly.

Others to try 'Clemson Spineless' is another All-America Selections winner from Clemson University, with tender green pods. 'Cow Horn' is a super-productive okra with 6–8 in. (15–20 cm) pods that stay tender. 'Stewart's Zeebest' is a Louisiana heirloom with tender, round pods.

'Burgundy' okra can easily do double duty in the flower garden, but do not forget to harvest its succulent pods.

ONION FAMILY

How different our kitchens would be without onions and garlic. They season and perfume so many wonderful dishes, that it is second nature to reach for them whenever we start to cook. They can also be used to create some fantastic dishes in their own right. Roasted garlic on warm bread can be served as a starter or as an entire meal. Onions and leeks shine as the stars of soups. And grilled onions caramelize so perfectly that you would think they were a dessert. All this lusciousness, and they even help to deter pests in the garden.

This is a subtle garlic used for seasoning and enhancing rather than overpowering other flavors. This beautiful softneck, artichoke type garlic is the winner of several taste tests and one of the most productive garlic varieties planted in home gardens. It is named after Inchelium, Washington, where it was discovered on the Colville Indian Reservation.

Bulbs are large and dense, with 8–20 cloves in concentric, enclosing circles, similar to the arrangements of artichoke bracts. They store well and get stronger in flavor as they age.

Garlic 'Inchelium Red'
Softneck garlic
Allium sativum var. *sativum*

Flavor 'Inchelium Red' has a mild heat and a garlic flavor that starts off subtle, but tends to fill out and permeate as you continue eating, enveloping the dish without overwhelming it.

Growing notes 'Inchelium Red' can be planted in late winter through early spring in frost-free climates, because it requires only minimal chilling. It should go in the ground about the same time you plant flowering bulbs. Softneck garlics do not overwinter well in cold climates. You can try planting them in the spring, but the bulbs will not grow as large.

Prepare a loose, well-drained soil bed with either lots of compost or a balanced fertilizer tilled in. Separate the garlic bulbs into individual cloves and plant the largest cloves with the pointed end facing up. Water well.

After the plants resume growing, apply more fertilizer or compost. Feed them again in 3–4 weeks. Keep your garlic well watered and remove weeds throughout the growing season.

How to harvest When the bottom leaves start to brown, test the garlic for ripeness. The only sure way to determine whether garlic is ready is to dig a bulb and slice it open. If the cloves are filled out, you can dig up the rest. Brush off excess soil and allow the bulbs to cure in a cool, dry, shady spot for 3–4 weeks before storing. 'Inchelium Red' is an excellent storing garlic, and its long, pliable necks can be braided and hung.

Tip Cloves of your own garlic, saved to plant the following season, will adapt to your growing conditions and become better and better over time.

Others to try 'Chet's Italian Red' is another Washington state heirloom with a mild flavor that is pleasant even raw. 'Lorz Italian' has large bulbs with a strong flavor and heat. 'Red Toch' is an excellent keeper with a flavor mild enough to eat raw, but rich enough to satisfy.

Exposure:
Full sun

Ideal soil temperature:
60°F (16°C)

Planting depth:
1–2 in. (2.5–5 cm)

Days to germination:
Depends on weather conditions

Spacing:
4–6 in. (10–15 cm)

Days to maturity:
150–240 days

'Inchelium Red' is a subtle garlic, but its flavor grows stronger as it ages.

'Spanish Roja' is a garlic lover's delight; the strong garlic aroma enhances and intensifies its flavor. It is a Rocambole hardneck garlic with a thin, parchmentlike skin that is easily peeled but does not store well.

Greek immigrants brought 'Spanish Roja' to Oregon sometime before 1900, and you may still see it listed as 'Greek' or 'Greek Blue'. The plants sport curling scapes with edible topsets known as bulbils, which also offer delicious garlic flavor.

AROMATIC

Garlic 'Spanish Roja'
Spanish red hardneck garlic
Allium sativum var. *ophioscorodon*

Flavor The flavor of 'Spanish Roja' is a fiery, warm spice. This garlic might "burn" your tongue when eaten raw, but it can be used to add a zesty bite to dishes, and the strong fragrance deepens the garlic flavor.

Growing notes 'Spanish Roja' cannot be planted in the spring because it needs 6–8 weeks of cold, 40°F (5°C) weather for the bulbs to develop. Gardeners in Zones 7 and colder can plant garlic any time after the first killing frost until the ground freezes. It is unreliable in frost-free areas.

Prepare a loose, well-drained

Exposure:
Full sun

Ideal soil temperature:
60°F (16°C)

Planting depth:
1–2 in. (2.5–5 cm)

Days to germination:
Depends on weather conditions

Spacing:
4–6 in. (10–15 cm)

Days to maturity:
150–200 days

The spicy, brownish-red cloves of 'Spanish Roja' add a zesty bite to dishes.

soil bed, with either lots of compost or a balanced fertilizer tilled in. Separate the garlic bulbs into individual cloves and plant the largest cloves with the pointed end facing up. Water well. If the temperature warms, you may see top growth in the fall. It will die back with a frost and resume growing in the spring.

Gardeners in areas with cold winters can apply a layer of mulch after the bulbs have been in the ground for about 4 weeks. This will prevent freezing and thawing of the soil, which could cause premature growth of the plants. You can remove the mulch in the spring or keep it there to conserve water. Apply more fertilizer or compost when new shoots appear and again in 3–4 weeks. Keep your garlic well watered and remove any weeds throughout the growing season. When the edible top scapes appear, cut them back to soil level to direct the plant's energy into the bulb.

How to harvest When the bottom leaves start to brown, test the garlic. The only way to determine whether garlic is ready is to dig a bulb and slice it open. If the cloves are filled out, you can dig up the rest. Brush off excess soil and allow the bulbs to cure in a cool, dry, shady spot for 3–4 weeks before storing.

Tip Garlic rarely sets seed and is always grown from cloves, so virtually all garlic is heirloom.

Others to try 'Chesnok Red' produces large bulbs that are great for roasting. 'Georgian Fire' has a strong but pleasant heat. 'Music' is an Italian variety with a sweeter heat.

Trust the land of vichyssoise to know its leeks. 'Carentan' is a chunky French heirloom that dates back to about 1885. It is extremely tender, with lovely, thick, blue-green leaves and succulent white stems that can grow to 2 in. (5 cm) in diameter and 8 in. (20 cm) long.

This late-season variety stands up to cold weather. You can even harvest it in winter if the ground is not frozen, although you might prefer it even more when it is a young, baby leek.

Exposure:	Full sun
Ideal soil temperature:	60–70°F (16–21°C)
Planting depth:	1/4 in. (0.6 cm)
Days to germination:	10–14 days
Spacing:	4–6 in. (10–15 cm)
Days to maturity:	100–130 days

Leek 'Carentan'
Carentan winter leek
Allium ampeloprasum var. *porrum*

Flavor 'Carentan' has a sweet, delicate herbal pungency that is closer to that of a shallot than a giant onion. The slightest cooking turns the leaves into a tender, buttery green. They add a gentle oniony warmth to dishes and are best used on their own, sautéed or roasted.

Growing notes Start seed indoors, 8–12 weeks before the last expected frost date. Gardeners in frost-free climates can plant 'Carentan' in the fall for a spring harvest.

Leeks require a long growing

season, and rich, fertile soil gets them off to a good start. Prepare a loose, well-drained soil bed and amend it with a large amount of compost or manure, several weeks before planting.

Start seeds indoors in flats. When the seedlings reach 3 in. (7.5 cm) tall, transplant them into individual cells. They are tiny, so do this carefully. Allow them to grow on inside, and after about 10 weeks from seeding, they should be the size of a pencil and ready to transplant outdoors.

Place the seedlings in a hole or trench about 6 in. (15 cm) deep. Fill loosely, leaving only a couple of inches (5 cm) of the plant above the soil line. After another 4–6 weeks, you can pile soil around the plants or mulch them with straw to start blanching the leaves. This will give you more tender white leaves and encourage the plants to grow taller. Add blanching material two more times before the fall.

How to harvest Harvest leeks in the fall by loosening the soil with a garden fork and lifting out the plants by the roots. Leave unneeded plants in the ground and mulch them to prevent heaving after the first hard frost.

Tip Leeks sweeten with a touch of frost. For easier winter harvesting in cold climates, grow leeks in a cold frame.

Others to try 'Blue Solaise' is a great choice for areas with a short season, with leaves that turn violet in the cold. 'Giant Musselburgh' is a large, mild, Scottish heirloom that is widely adaptable. 'Prizetaker' is an English heirloom with tall, tender, white stalks.

'Carentan' is a beautiful, edible, blue-green delight.

Exposure:
Partial shade to shade

Ideal soil temperature:
60-75°F (16-24°C)

Planting depth:
Sow seeds on top of soil and press in

Spacing:
Broadcast freely

Days to germination:
6-18 months

Days to maturity:
2-3 years

Sometimes the most wonderful things are hidden in plain sight. Ramps are a small, mellow leek that grows wild in American woodlands from Canada to South Carolina. Bright, warm green ramps put in a brief appearance each spring and then disappear for the rest of the year. You have to wonder why something so delicious should be relegated to the wild.

Ramp bulbs are similar to scallions, and the leaves could easily be mistaken for lily of the valley. You can find them growing in bunches in early spring. By midsummer, ramps will have gone to flower.

The town of Richwood, West Virginia, has been holding the Feast of the Ramson in April for more than 70 years. Chicago pays a more dubious tribute to ramps. According to the *Encyclopedia of Chicago*, the name "Chicago" is from a Native American word for striped skunk, which was also the word used for ramps.

Leek, ramps
Ramson, wild leek
Allium tricoccum

Be patient and wait for ramps; their numbers should increase every year.

Flavor Ramps smell like garlic but taste like wild onions. They are not as sulfuric or pungent as cultivated garlic and onions, but they do have a pleasant kick when freshly harvested. The longer they are stored, the less flavorful they become.

The green tops are also edible, with a milder kick. You can

use ramps the same way you would use spring onions, but to enjoy their unique flavor, try roasting or pickling them, or chop them fresh and use them to finish off a dish.

Growing notes Ramps can be transplanted throughout spring. You can start ramps from seed in late summer to early fall. The seeds require a warm, moist period followed by a period of cold to break dormancy. Depending on the weather, seed germination may take more than a year.

It is still rare to find ramps for sale, and if you do, they are usually sold as potted transplants in the spring. Find a shady spot where they can spread out, and plant them so that the bulbs are just below the soil surface.

A moist, humus-rich soil, much like that of their native forest, is your best bet for success. Mulch them lightly and keep them cool and watered. Be sure to mark the spot because they can be ephemeral.

How to harvest Ramp plants can take 2–3 years to fill out enough to start harvesting. You can pull a few bulbs when the plants leaf out in midspring, but be sure to leave some in the ground to multiply.

Tip Foraging for ramps and removing whole plants has caused some areas to become overharvested and depleted. Many public sites have ramp harvesting restrictions, and some now require harvest permits.

'Long Red Florence' offers much to recommend it, including the fact that you can grow a lot of them in a small space. These spindle-shaped onions are mildly spicy but lively and delicious for fresh eating. Each bulb has a distinctive shape that garners it many names, such as 'Red Torpedo' and 'Italian Red Bottle'. The 4 in. (10 cm) long, narrow bulbs can be grown in tighter spaces than globe onions. 'Long Red Florence' earned the Royal Horticultural Society's Award of Garden Merit.

Onion 'Long Red Florence'

Torpedo onion
Allium cepa

Flavor These onions are full of warm, spicy onion flavor with almost no heat. You can slice them into crunchy half moons with no fear of tearing up and toss them into a salad. They can be used in place of regular globe onions,

but their gentle onion flavor is somewhat lost in cooking; they do make nice pickles, however.

Growing notes If you are direct sowing, plant seeds in a loose, well-drained, and fertile soil, 1–3 in. (2.5–7.5 cm) apart and thin as necessary. For a head start, you can start seed indoors 4–6 weeks before the last expected frost date. When the seedlings are 2–3 in. (5–7.5 cm)

Exposure:
Full sun

Ideal soil temperature:
70–75°F (21–24°C)

Planting depth:
1/4 in. (0.6 cm)

Days to germination:
10–14 days

Spacing:
3–4 in. (7.5–10 cm)

Days to maturity:
95–120 days

'Long Red Florence' torpedo onions are spicy but easy on the heat.

tall, feed them a dilute solution of kelp or seaweed emulsion and clip off a third of the leaves. Keep them indoors until any danger of frost has passed, and then harden them off for 5–7 days by moving them outside during the day and indoors at night. Gardeners in frost-free climates can sow seed in the fall for a spring harvest.

Be gentle when transplanting the delicate seedlings. Onions are shallow rooted and need lots of water to plump up. Keep the beds free of weeds and mulch to conserve water. Feed them once a month while the leaves are growing.

Onions require a long growing season, especially when grown from seed. Onion growth is also affected by the amount of sunlight the plants receive, and onion varieties are categorized according to the optimum day length for their growth. 'Long Red Florence' is considered an intermediate day length onion, meaning it requires about 12–14 hours of sunlight before it will begin to form bulbs. Intermediate day length onions are the most adaptable.

How to harvest 'Long Red Florence' signals it is ready to harvest when most of its leaves have fallen over. Gently dig and lift the bulbs and allow them to cure in a cool, dry spot. Bending the remaining leaves when they first start to fall will speed up harvest time and encourage the onions to form protective outer skins.

Others to try 'Amish Bottle' is a mild, white variety. 'Rossa di Milano' is a spicier red variety with a real kick. 'Tropeana Lunga' is very sweet.

Whatever you call it, this is one odd onion. Not the usual bulb onions, these perennial onions form small bulbils at the ends of long, hollow stalks. 'McCullar's White Topset' will sprout while still attached to the original plant, with a Medusa-like appearance. After they get top heavy, they fall over and replant themselves, as though they are walking along the garden. Topsetting onions add interest, humor, and a bit of sculpture to the garden. They spread themselves freely—and thank goodness for that, or you might hesitate to eat them. They are rapid spreaders and offer a steady supply of greens all summer.

Many people are unsure about what to do with topsetting onions, but these wonderful little oddities offer lots of options. You can eat the small bulbils, the greens, and even the flower stalks. The small bulbils form in late summer and into the fall.

Exposure:
Full sun to partial shade

Ideal soil temperature:
60–75°F (16–24°C)

Planting depth:
1/2–1 in. (1–2.5 cm)

Days to germination:
7–14 days

Spacing: Plant:
6–9 in. (15–22.5 cm)

Days to maturity:
Matures the following fall

Onion 'McCullar's White Topset'

Walking onion, tree onion, topsetting onion
Allium cepa var. *proliferum*

Flavor Bulbils are more intensely pungent than traditional onions, and they are a bit hotter than bulb onions but more flavorful than some of the red varieties. They are great for cooking, pickling, and flavoring vinegars but are probably too strong for most people to enjoy fresh. The greens taste much like scallions or chives and can be used to zip up dishes. The hollow stems

You can eat the small bulbils, the greens, and even the flower stalks of 'McCullar's White Topset' onions.

can be harvested with the bulbils and make a nice treat when stuffed with cheese.

Growing notes The most difficult part of growing 'McCullar's White Topset' is finding them in the first place. If a friend is willing to share some bulbs, you are ahead of the game. Young plants can be transplanted in the spring, and bulbils are planted throughout the fall. Fall is the best time to get them established. When the flower stalks start to turn brown, you can remove the bulbils; choose the largest ones for replanting. Plant them shallowly and keep them well watered until they sprout.

A well-drained, sandy loam amended with compost is ideal for growing these onions. All onions are heavy feeders and benefit from additional side dressings with compost or composted manure throughout the season.

How to harvest You can cut the greens any time to use as scallions or chives. The topsets form in the late summer and will eventually fall off the plant. Harvest the underground bulb by digging up the entire plant.

Tip To prevent walking onions from walking their way through your vegetable garden, harvest all the topsets before they reach the ground.

Others to try 'Catawissa' has reddish-brown bulbils. 'Fleener's Topset' produces topsets plus a small underground bulb. 'Moritz' produces purplish-red, slightly larger topsets.

Exposure:	Full sun
Ideal soil temperature:	70–75°F (21–24°C)
Planting depth:	1/4 in. (0.6 cm)
Days to germination:	10–14 days
Spacing:	3–4 in. (7.5–10 cm)
Days to maturity:	80–90 days

'Red Marble Cipollini' (pronounced chip-oh-lee-nee) is a crunchy little flattened onion with a sweetness that falls between shallots and Vidalia onions. Their unusual beauty is part of the pleasure of cooking with them. The deep red color is carried throughout the 1-by-1½ in. (2.5-by-4 cm) bulb.

Cipollina means small onion in Italian. Most of the varieties available in North America can all be traced back to an onion brought over in the 19th century, now called 'Italian Red'. 'Red Marble Cipollini' is solid and dense and keeps for months.

Onion 'Red Marble Cipollini'

Sweet onion
Allium cepa

Flavor This Italian version of a pearl onion is meant to be eaten whole. Their high sugar content gives them a refreshing sweetness. Combine that with their firm texture and the complex earthy onion zest, and you have an onion that is perfect for roasting or glazing.

Growing notes This long day length onion needs 15–16 hours of daylight but produces harvestable onions in less time than

most others. It does best when it is grown during the summer in cooler regions, but gardeners in warmer climates can try planting in midsummer.

In Zones 7 and cooler, direct sow seeds in early spring, in a loose, well-drained, fertile soil, 1–3 in. (2.5–7.5 cm) apart, and then thin seedlings as necessary. For a head start, start seed indoors 4–6 weeks before the last expected frost date. When the seedlings are 2–3 in. (5–7.5 cm) tall, feed them a dilute solution of kelp or seaweed emulsion and clip their leaves by a third. After

any danger of frost has passed, harden them off for 5–7 days by moving them outside during the day and back indoors at night.

Grow 'Red Marble Cipollini' as you would regular sized onions. Start with a rich, loose, well-draining soil. Gently pull apart and transplant the delicate seedlings 2–3 in. (5–7.5 cm) apart.

Onions are shallow rooted and need lots of water to plump up. Keep beds free of weeds and mulch to conserve water. Feed them once a month while the leaves are growing.

How to harvest The leaves

'Red Marble Cipollini' onions are juicy and sweet enough to enjoy fresh.

will start to bend and fall when it is time to harvest. Their little bulb shoulders will also poke through the surface. Gently dig and lift the bulbs and allow them to cure in a cool, dry spot until the necks are completely dry and papery, which generally takes 2–4 weeks.

Tip Onions with thicker necks do not store well and should be eaten first.

Others to try 'Bianca di Maggio' is another small onion with a delicate flavor. 'Borettana' is firm and sweet and great for grilling. 'Red Creole' is a bulkier, spicy, red, short-day variety that is good for gardeners in warmer climates.

Potato onions are not as well known as their cousins, French shallots, but these lively morsels are far more versatile and easier to grow. After you plant a small mother bulb in either the fall or spring, the mother splits into many smaller bulbs that sprout tender, scallionlike leaves. 'Yellow Potato', which has been grown since prior to 1886, was originally prized for its green tops alone. They were called rare-ripes and were anticipated as a springtime treat.

Exposure:
Full sun

Ideal soil temperature:
70–75°F (21–24°C)

Planting depth:
1/2–1 in. (1–2.5 cm)

Days to germination:
10–14 days

Spacing:
6–8 in. (15–20 cm)

Days to maturity:
80–90 days (spring planted); 250 days (fall planted)

Onion 'Yellow Potato'

Potato onion, multiplier onion
Allium cepa var. *aggregatum*

Flavor Young 'Yellow Potato' onions have a mild sweetness comparable to Dutch shallots and are delicious when eaten raw or slightly cooked. As they age in storage, they develop a complex flavor of peppery bite and mellowed seasoning. They can be used as pearl onions or as a substitute for regular onions. Their one drawback is they are not the easiest onion to peel.

Growing notes Potato onions can be planted in fall or early spring. Barely cover the bulbs sets with 1/2–1 in. (1–2.5 cm) of soil. In Zones 7 and cooler, you will need to mulch them for winter with another 2–4 in. (5–10 cm) of soil. Add a 5–6 in. (12.5–15 cm) layer of straw after the first hard frost to prevent heaving. Remove all the mulch in the spring. Alternatively, plant the bulbs in spring with their pointed tips just peeking out at soil level. Multiplier onions form best when they are grown at the soil surface. The largest bulbs will result from sets planted in the fall.

Potato onions are grown much like garlic and other shallots. Start with a rich, well-draining soil. They grow shallowly and do not require a deeply tilled soil. Other than needing regular watering, the bulbs grow without fuss.

How to harvest Onions sprout in early spring and you can harvest some of the green tops, but leave enough to keep the plants growing. Jolly clusters of small, golden skinned onions

will push their way out of the soil to tell you they are ready for harvest. When the leaves start turning brown, you can dig the bulb clumps and allow them to cure in a cool, shady, dry spot. You should get 5–10 onions per set. 'Yellow Potato' is an excellent keeper that lasts throughout the winter.

Tip First-year bulbs will not grow large. As you save and replant bulbs, they will adapt to your garden and future generations will grow into chubby 3 in. (7.5 cm) orbs.

Others to try 'French Red' shallot is rich and spicy. 'Odetta's White' shallot is a tiny, mild heirloom from Kansas. 'Red Potato' is a milder, less robust multiplier variety.

Keep saving and replanting 'Yellow Potato' bulbs and you may never have to buy an onion again.

PARSNIPS, RUTABAGAS, SALSIFY, AND TURNIPS

Root vegetables are the mysteries of the vegetable garden. You plant the tiny seeds early, and then cross your fingers that all is going well beneath the soil surface, until fall, when those seeds have grown into underground treats. These four vegetables do not disappoint.

Although often lumped together because of their similarities in growing conditions and their shared homeliness, these root vegetables offer a lot of versatility and a depth of complex flavors. Sweet, nutty parsnips were considered so fine they were fit for the aristocracy of ancient Rome. Rutabagas, a turnip-cabbage cross, can be sharp and savory when raw or used as a creamy base for spices when cooked. If you have never heard of salsify, take a look at Thomas Jefferson's garden records and recipe books. He grew them with the same enthusiasm as carrots. Turnips were once as popular as tomatoes, with several dozen varieties being grown and circulated.

The trick to mastering this quartet in the garden is patience in the ground and a light hand in the kitchen. Cook them gently and savor their taste, and you will be delighted to make room for them in your garden.

All parsnips require an act of faith and patience. They are planted in the chill of spring and then grow, out of sight, until the chill returns in the fall. The thick roots will reach down 15 in. (38 cm) or more, with few side roots along their smooth skin.

Wild parsnips have been harvested and eaten since the Stone Age. They have been refined since then, but they can still seem a bit primeval. Although early English colonists brought parsnips to the New World, 'The Student' was not introduced until 1860. Surprisingly sweet and complex, 'The Student' is a delight. It remains one of the two most popularly grown parsnips in home gardens because it grows so reliably and the flavor is worth the wait.

Exposure:
Full sun

Ideal soil temperature:
70°F (21°C)

Seed Planting depth:
1/8–1/4 in. (0.3–0.6 cm)

Days to germination:
14–20 days

Spacing:
4–8 in. (10–20 cm)

Days to maturity:
95–125 days

'The Student' is a widely grown parsnip whose flavor is worth the wait.

Parsnip 'The Student'
Pastinaca sativa

Flavor 'The Student' was bred with flavor in mind, with a mellow sweetness. Sweet and earthy, starchy, and herbal, this parsnip can be savored like a fine wine. Do not be so quick to cream and mash the tubers. Simmer them gently and season them as a main course, with your finest olive oil, white wine, and herbs.

Growing notes 'The Student' is slow to sprout and takes its time to reach maturity. Direct seed in the garden as soon as the soil has warmed and dried out in the spring. Gardeners in frost-free areas can seed in late summer through fall. To ensure that the crop sweetens, time your seeding so that parsnips will be maturing during the coolest months. Mature parsnips can remain in the ground through the fall and even winter.

Parsnip seeds deteriorate with age, and the freshest seed will produce the longest roots. Seed goes dormant when dried and needs some time in the chilly spring soil to break dormancy, so get them in the ground as early as possible.

A deep, loose soil is essential for straight roots. The tiniest pebble can cause a turnip to fork or bend. After planting, lightly cover seed with sand; they can have difficulty breaking through caked soil. Sandy soil is heaven for parsnips, although some compost worked in will help keep them fed. Oddly, manure tends to cause forking.

After you prepare the soil and the seeds germinate, your job is simple: provide water, keep the area free of weeds, and have patience.

How to harvest Parsnip plants need to be exposed to frost to get the sweetest taste. You can start harvesting in late fall and leave them in the ground well into winter.

Tip Create a couple of raised beds in your garden and fill them with sandy soil. Then use them to alternate root crops each year, amending annually with compost.

Others to try 'Guernsey' is stocky and sweet. 'Harris Model' is a nice choice for a fall-planted crop to overwinter. 'Hollow Crown' is an old standard, with long, sweet roots.

Golden, chubby, sweet, and misunderstood, if ever a vegetable were in need of a public relations campaign, it is the rutabaga. Food writer Elizabeth Schneider claims that if the rutabaga had been given a name such as "golden globe" it would be hailed as queen of the root crops. 'Laurentian' rutabaga is creamy, sweet, and multitalented; even the greens have a kalelike lusciousness. Rutabagas are thought to be a cross between cabbage and turnips, and 'Laurentian' definitely gets its greens from the cabbage side of the family.

This rutabaga is a beautiful pale, golden globe, blushing reddish-purple where it pokes its shoulders above ground. Rich in beta-carotene and low in calories, it is an improved version of the old standard 'American Purple Top', developed in the early 20th century. 'Laurentian' is at home in dishes from pickles to pies.

Rutabaga 'Laurentian'
Swede turnip
Brassica (Napobrassica Group)

Flavor 'Laurentian' has a peppery sweet taste that makes it a versatile vegetable. When eaten raw, it has a bold, sharp flavor similar to that of its cousin, the cabbage. Cooking softens the flavor and brings out the sweetness. Roasting in particular will highlight its firm texture and caramelize its sugars. It makes a surreptitious substitute for potatoes, but do not boil it to a mush.

Growing notes 'Laurentian' can be seeded or transplanted in late spring after the last frost or in early summer for a fall harvest. In frost-free climates, it is a winter crop that can be seeded or transplanted in the fall.

As a root crop, rutabagas need a loose, well-draining soil. Soil need not be overly rich, but some potash and bone meal will help the roots plump up. Too much high-nitrogen organic matter can cause deformed roots. No additional feeding should be

necessary. Keep the area free of weeds and well watered, especially during hot spells. Prolonged high temperatures and drought will cause the roots to crack.

Row covers will keep the greens pristine. Root maggots can spoil the roots, and rotating your rutabagas to different areas of the garden each year will help offset their damage.

How to harvest Harvest rutabagas in the fall, after a frost has sweetened them. They should be about 3–6 in. (7.5–15 cm) in diameter. Although 'Laurentian' stores well after harvesting, you can always leave these rutabagas in the ground, protected with a straw mulch, and harvest as needed.

Tip 'Laurentian' can be difficult to peel. Microwave the root for about 1 minute to make it easier to cut into quarters, and then peel the pieces.

Others to try 'American Purple Top' is similar in flavor and texture but slightly smaller and more tapered. 'Nadmorska' is a

Exposure:
Full sun to partial shade

Ideal soil temperature:
65–70°F (18–21°C)

Planting depth:
1/2–1 in. (1–2.5 cm)

Spacing:
6–8 in. (15–20 cm)

Days to germination:
7–15 days

Days to maturity:
90–120 days

The size of a fresh rutabaga depends on the length of the growing season.

large, sweet Lithuanian variety that matures early. 'Wilhelmsburger' is a vigorous German variety with tender, gold flesh and a green top.

Subtle, mellow, and undeniably unique, salsify is a vegetable to season gently. It is called the oyster plant because its flavor can evoke that of oysters, especially when salsify is simmered in soup. Peeling salsify's thin roots can make it as difficult to prepare as oysters, which is why 'Mammoth Sandwich Island' is so wonderful: its longer, plumper roots mean less work and more wonderful flavor.

'Mammoth Sandwich Island' has been around since the 1800s, and if any variety can make salsify popular, this is the one. The roots are fairly uniform, so it is easy to peel and get to the sweet and tender, creamy white flesh. It also retains these wonderful qualities in storage, so you can enjoy it throughout the winter.

Salsify 'Mammoth Sandwich Island'

Oyster plant
Tragopogon porrifolius

Flavor With a texture similar to a tender carrot that becomes creamy when cooked, this salsify is earthy, yet fresh, and reminiscent of cooked oysters. The tender texture gets silkier with cooking. Try a touch of lemon juice to help enhance the complex flavors.

Growing notes Salsify does not transplant well. For the best results, direct sow seeds in the garden. If you must start them indoors, sow in peat or paper pots to ease transplant shock. Keep the soil moist until they germinate.

In Zones 4–8, start seed outdoors, about two weeks before the last expected frost. Gardeners in Zones 8 and warmer can plant seed in the fall, to grow plants ready to harvest the following spring, but the flavor is not equal to roots that mature in the fall.

Because salsify roots can be quite long, loosen the soil well before planting. You may even want to dig holes about 12. in. (30.5 cm) deep and refill them with compost to help ease the way for the roots. Salsify likes a rich soil, but too much nitrogen will cause the roots to split.

The first sprouts look like short, dark twigs and are easy to miss. They will eventually start resembling blades of grass, so mark the area to keep from accidentally weeding out your plants.

Regular watering will keep the salsify roots tender. They also appreciate some shade and mulch during hot spells. Side dressing with more compost midseason will keep them going into the fall.

Few things bother salsify plants, except competition from nearby weeds.

How to harvest Salsify is ready to harvest about 120 days after planting. Use care when digging the roots: broken roots do not store well. Spring-planted salsify can be dug in late fall or can overwinter in the ground. Like parsnips, the flavor sweetens with a touch of frost,

Exposure:
Full sun to partial shade

Ideal soil temperature:
40–55°F (5–13°C)

Planting depth:
1 in. (2.5 cm)

Days to germination:
7–21 days

Spacing:
3–4 in. (7.5–10 cm)

Days to maturity:
120–150 days

You can certainly see how salsify could be mistaken for a clump of grass and weeded out of the vegetable garden.

but the unique oyster flavor is diminished.

Tip Remove the leaves before refrigerating to retain the root's sweetness.

Others to try 'Fiore Blu'

('Blue Flower') has darker roots with a deep, musky flavor. Scorzonera hispanica, or black salsify, is related to salsify, with a more uniform shape and a similar oyster flavor.

'Golden Ball' turnip has a flavor as unexpected as its appearance. It is mild, sweet, and nutty with a texture so tender, you can eat it fresh from the garden. In fact, it also goes by the name 'Orange Jelly' because it is so melt-in-your-mouth tender. Although this turnip has been grown since the mid-1800s, you will not often find it outside the backyard vegetable patch.

'Golden Ball' is not really golden. It is a pale orange on the outside with yellow flesh. The color varies greatly depending on the soil, but the flavor and tenderness remain constant.

Turnip 'Golden Ball'
'Orange Jelly'
Brassica rapa

Flavor 'Golden Ball' is smooth and sweet, with an unusual almond aftertaste. It can add an unexpected flavor to salads. It is especially creamy when cooked and mashed. Its sweetness also makes it a great choice for roasting.

Growing notes Direct seed in the garden as soon as the soil can be worked. Succession planting every 3–4 weeks until midsummer will ensure a steady harvest. Gardeners in frost-free climates will have better luck planting 'Golden Ball' in the fall and through winter.

Turnips are at their best in cool weather. Although they will tolerate most soils, a loose, rich soil will allow the roots to grow quickly and stay tender. Keep the seed moist until it germinates, and then provide water throughout the growing season. It tends to bolt to seed in warmer weather and in dry soil.

Because 'Golden Ball' matures fairly quickly, no fertilizer is needed in rich soil. Side dress with compost if soil is lean. Keep the area free of weeds to avoid competition for nutrients.

Plants can be thinned by cutting off the tops. This will prevent nearby turnips from being disturbed. You can eat the tiny turnip greens.

The two most serious pests are root maggots and voles. To deter root maggots, do not plant 'Golden Ball' where turnips, rutabagas, or radishes were grown the preceding year. Voles love turnips, so using traps or some type of ground barriers may be necessary.

How to harvest Harvest when the bulbs are about 3 in. (7.5 cm) in diameter. Larger bulbs can be pithy. 'Golden Ball' gets sweeter with a frost and will store well into spring.

Tip If you do not have a cold place to store harvested turnips, they can be blanched and frozen for use throughout the winter.

Others to try 'Gilfeather' is a white, egg-shaped turnip with a sweet, mild flavor. 'Shogoin' is a small, white Japanese variety that stays tender in hot weather. 'White Egg' is a small, white turnip that is mild and creamy.

Exposure:
Full sun

Ideal soil temperature:
65–70°F (18–21°C)

Seed Planting depth:
1/8–1/4 in. (0.3–0.6 cm)

Days to germination:
14–20 days

Spacing:
4–8 in. (10–20 cm)

Days to maturity:
50–70 days

(Following page) 'Golden Ball' is small and delicate, with a dense and tender texture.

PARSNIPS, RUTABAGAS, SALSIFY, AND TURNIPS

PEPPERS

Hot or sweet, long or blunt, vibrantly colored or dusky and dark, peppers add zest to a meal any way you slice them. Cultivated some 2000 years ago in South and Central America, peppers were first introduced to Europe by Christopher Columbus. They have traveled the world and are now integral parts of almost every cuisine.

Peppers are heat-loving perennial plants that thrive during the sultry days of summer, blossoming and ripening faster and faster as the days heat up. They require minimal care, and the flavors and heat actually improve with a little neglect. Eat them fresh or sautéed and freeze some for a reminder of summer after those sultry days have passed.

Glowing 'Bulgarian Carrot' peppers do not last long on the plant, so harvest them soon after they ripen.

Shiny, smooth, and day-glow orange, these peppers look juicy and sweet, like baby carrots. But do not be fooled: they pack a decidedly spirited kick. Supposedly smuggled out of communist Bulgaria, 'Bulgarian Carrot' is far more cheerful than sinister, gradually changing from bright green, to yellow, to carrot orange. The 2 ft. (0.6 m) plants are high yielding and strikingly ornamental.

Exposure:
Full sun

Ideal soil temperature:
75-85°F (24-30°C)

Planting depth:
1-1½ in. (2.5-4 cm)

Days to germination:
10-14 days

Spacing:
18-24 in. (45.5-61 cm)

Days to maturity:
70-80 days

Pepper 'Bulgarian Carrot'

Capsicum annuum

Flavor You may sense a subliminal carrot flavor when you taste 'Bulgarian Carrot', and its spicy heat is enhanced by an acidic, citrus aftertaste. The flesh is juicy, but the skin can be tough to chew, so you can chop these peppers for use in salsas or as pickles. If you are adventurous, you can use them to make spicy jelly.

Growing notes Start seed indoors, 8-12 weeks before your last expected frost date. Do not transplant seedlings outdoors until all danger of frost has

passed. Gardeners in Zones 8 and warmer can transplant a second crop of peppers in midsummer to mature in the fall.

Pepper seeds germinate more quickly if the potting medium is kept warm. They will grow slowly at first and require warm temperatures and bright light. Feed them with a half-strength, soluble liquid fertilizer after their true leaves appear. Keep the plants indoors until nighttime temperatures remain above 50°F (10°C), and then slowly start to harden them off, leaving them outdoors for increasing amounts of time over a 7-10 day period.

Transplant seedlings about

1 in. (2.5 cm) deeper than they were in their pots. Peppers prefer a slightly rich, well-drained soil. Your biggest hurdle when growing peppers is extreme weather. Excessive heat, cold, rain, or drought will cause the peppers to drop their flowers. Be patient, however, because they will recover.

'Bulgarian Carrot' tends to start producing peppers from the bottom of the plant up. Staking will keep it standing upright and allow plenty of sunshine to reach inner leaves and fruit.

How to harvest You can harvest these peppers at any color stage, but the fullest flavor develops when the peppers are bright orange and glossy. Cut the peppers from the plant to avoid damaging the stem.

Tip The heat of hot peppers varies with the quality of the soil. Stressing the plants slightly, by crowding or withholding water, intensifies their heat.

Others to try 'Black Hungarian' is mildly spicy, with deep purple-black coloring and thick walls. 'Bulgarian Apple' is a stocky, spicy aji type pepper (aji peppers are used in the South American condiment chimichurri) that matures to beet red. 'Hungarian Hot Wax' is a short season, pale yellow pepper with lots of heat; it makes great pickles.

If the name and the flaming golden coloring do not offer enough hint of what is in store, one bite will tell you all you need to know. 'Fatali' has been listed as the sixth hottest pepper tested, at a sizzling 125,000–325,000 Scoville heat units. These are so hot, you can smell their heat.

The 3 in. (7.5 cm) gnarled fruits look similar to long habaneros, with thin, crisp walls and few seeds. This central African native needs a long, hot growing season. The colorful 2 ft. (0.6 m) plants do well in pots and can be brought inside during the winter months.

SPICY

Pepper 'Fatali'
Capsicum chinense

Flavor Lurking under the heat is a definite lemony citrus flavor. Unless you are a true hot pepper worshipper or have an asbestos stomach, these peppers are best used as a seasoning, not as a popper. They are unparalleled for making hot sauces and for perking up salsas and cooked dishes—when used in small amounts.

Growing notes Start seed indoors 8–12 weeks before your last expected frost date. Do not transplant outdoors until all danger of frost has passed. Gardeners in Zones 8 and warmer can transplant a second crop in midsummer to harvest into fall.

Exposure:
Full sun

Ideal soil temperature:
75–85°F (24–30°C)

Planting depth:
1–1½ in. (2.5–4 cm)

Days to germination:
10–30 days

Spacing:
18–24 in. (45.5–61 cm)

Days to maturity:
80–90 days

'Fatali' seeds can take weeks to germinate. Heating the seeds flats from the bottom will speed germination a bit. Give the seedlings plenty of bright light and keep them warm. Feed with a half-strength liquid fertilizer after their true leaves appear. When nighttime temperatures remain above 50°F (10°C), you can slowly start to harden them off, leaving them outdoors for increasing amounts of time over a 7–10 day period. Do not rush this adjustment period. 'Fatali' will not start flowering and producing until midsummer, so take your time.

Transplant seedlings about 1 in. (2.5 cm) deeper than they were in their pots. 'Fatali' plants are stubbornly slow to grow and kick into gear only when summer heats up. They will not produce without lots of heat and sunshine, and they seem to thrive on neglect and produce higher, hotter yields if you allow them to become slightly drought-stressed. The plants grow bushy, and the weight of the hollow peppers does not pull them down. You will not need to stake them, but give them room to spread out.

How to harvest Harvest these peppers at any color stage. The young, green peppers are milder, but not by much. Cut the peppers from the plant to avoid damaging the stem.

Tips Handle these peppers with care; avoid contact with eyes and sensitive skin. If you bite into a pepper that is too hot to handle, do not reach for water. Dairy products, such as milk and yogurt, and bread are better choices for helping to dissipate the heat in your mouth.

Others to try 'Caribbean Red' habanero is a high-heat pepper with a smoky citrus taste. 'Red Mushroom' is a flattened dollop-shaped pepper with fruity heat. 'Tobago Seasoning' is a spicy flavored but mildly hot pepper.

'Friggitello' are not your usual hot peppers—in fact, they are not hot at all. These peppers lend themselves to pickling, but even unadorned, they are crisp, with a very pleasant warm bitterness. In Italy, they are considered sweet peppers.

You may be familiar with pepperoncini as the pickled salad bar garnish, but these are often the Greek variety, which tend to be sweeter. The Italian 'Friggitello' is most associated with Tuscany.

Exposure:
Full sun

Ideal soil temperature:
75–85°F (24–30°C)

Planting depth:
1–1¹/₂ in. (2.5–4 cm)

Days to germination:
10–14 days

Spacing:
18 in. (45.5 cm)

Days to maturity:
70–80 days

'Friggitello' is a crunchy, thin-walled pepper that has more flavor than heat, even when eaten fresh.

Pepper 'Friggitello'
Pepperoncini
Capsicum annuum

Flavor Forget what you know about pickled pepperoncini. The complex flavors of 'Friggitello' are best appreciated when it is freshly picked. Its crinkly, pale green skin makes a lovely, juicy crunch and releases a head-filling pepper flavor with all the zest you could want, without the heat. It tends toward warmth, rather than heat, and the full flavor resonates in your mouth. These are wonderful peppers for sautéing. They are perfect for stuffing and popping; surprisingly good mixed with fruits, in salads, and in compotes; and great on sandwiches.

Growing notes Start seed indoors 8–12 weeks before your last frost date, and set out transplants after all danger of frost has passed. A midsummer planting is possible for gardeners in Zones 8 and warmer, to harvest into fall.

As with most peppers, 'Friggitello' seeds are slow to germinate. Adding some bottom heat will speed germination. Once they are up, the seedlings need plenty of bright light and warmth. Feed with a half-strength liquid fertilizer after their true leaves appear. When nighttime temperatures remain above 50°F (10°C), you can slowly start to harden them off, leaving them outdoors for increasing amounts of time over a 7–10 day period.

Transplant seedlings when they are 5–7 weeks old and just filling out. Plant them about 1 in. (2.5 cm) deeper than they were in their pots. Peppers like a slightly rich, well-drained soil, but do not overfertilize. Soil that is too rich will result in many leaves and few peppers.

'Friggitello' can set large amounts of fruit, and staking is required or the plant will become top heavy and tip over.

How to harvest Start to harvest when the peppers reach 2–3 in. (5–7.5 cm) long. No need to wait for them to change color.

Tip A good rule of thumb is to plant peppers when the bearded iris bloom.

Others to try 'Beaver Dam' is warm and crunchy, with a sweet aftertaste. 'Hinkel-hatz' (chicken heart) is a Pennsylvania Dutch heirloom that is great for pickling. 'Poblano' is a traditional Mexican pepper (called ancho when dried), with a rich, savory flavor.

Long, slender, crunchy, and bursting with peppery spirit, 'Jimmy Nardello' is everything a great sweet pepper should be. The background story of this pepper is an heirloom classic. Giuseppe and Angela Nardiello (the family later dropped the i from the name) grew this pepper every year in their garden in southern Italy's Basilicata region. Naturally, they brought seed with them when they set off for America in 1887, reportedly sewn into the hem of Angela's dress. They named this pepper after their fourth son Jimmy, and he continued to plant these seeds throughout his lifetime, eventually passing the seeds and the story along to Seed Savers Exchange. Both the story and the peppers have been charming heirloom gardeners since.

This is the perfect pepper for cooking. The thin skins sauté to a delectable sweetness. Slow Food USA has included 'Jimmy Nardello' in its Ark of Taste, a catalog of more than 200 delicious foods that are in danger of extinction.

Exposure:
Full sun

Ideal soil temperature:
75-85°F (24-30°C)

Planting depth:
1-1½ in. (2.5-4 cm)

Days to germination:
10-14 days

Spacing:
18-24 in. (45.5-61 cm)

Days to maturity:
80-90 days

Red-lacquered 'Jimmy Nardello' peppers twist and turn as they grow.

Pepper 'Jimmy Nardello'

Sweet pepper
Capsicum annuum

Flavor 'Jimmy Nardello' has a fresh, fruity, cherrylike sweetness. The peppers' crispy, juicy, thin walls hold their firmness when cooked. This is the perfect pepper for sautéing, but it is also wonderful eaten fresh or stuffed.

Growing notes Start seed indoors about 8-10 weeks before your transplant date. Transplants can go outside after all danger of frost has passed. Gardeners in frost-free climates can start a second crop in midsummer to harvest in the fall.

Transplant seedlings into individual pots when they are about 2 in. (5 cm) tall. When all danger of frost has passed, you can harden them off by increasing the amount of time they spend outside over a 1-2 week period. Seedlings can be planted in the garden slightly deeper than they were in their pots.

Peppers like a moderately rich, well-drained soil. 'Jimmy Nardello' starts setting early and needs regular watering to keep growing and producing. A dose of fertilizer or compost midseason will also give the plants a boost. Sweet peppers tend to grow a bit more quickly than hot peppers, and they grow into larger plants, so staking is required.

How to harvest 'Jimmy Nardello' is sweetest and juiciest when it is glowing red, but you can harvest green peppers if you are after a stronger, more peppery flavor.

Others to try 'Chervena Chushka' is a super sweet and mild Bulgarian pepper that is great for cooking. 'Healthy' is a crunchy, sweet Russian pepper that bears early and performs well in cool weather. 'Marconi' is a sweet elongated Italian bell pepper.

"Cute" and "cheerful" may not be common adjectives in the vegetable aisle, but 'Red Mini Bell' peppers are undeniably pert. The size of a large walnut, each is the perfect party pepper and as juicy as its cherry color would suggest, with a pleasantly sweet, peppery zest.

Unlike "baby" squash and other so-called baby vegetables, 'Red Mini Bell' is fully mature at 1½–2 in. (4–5 cm). This is not the hot, vinegary pepper in the pickle jar. It is a sweet, luscious fruit in a bold red package, waiting to be popped into your mouth.

'Red Mini Bell' seed was given to Seed Savers Exchange by Lucina Cress, whose family had been growing them for years. They would stuff them with cabbage, pickle and can the lot, and sell them in pint jars at church fundraisers.

Pepper 'Red Mini Bell'
Mini bell pepper
Capsicum annuum

Flavor 'Red Mini Bell' has opulent, thick walls of glowing red flesh—each is a little jewel. They pack the full spiciness of peppers with a rich sweetness that combine to ambrosia in a miniature package. Fresh, stuffed, pickled, or tossed whole into stir-fries, they are as addictive as potato chips.

Growing notes Start seed indoors about 8 weeks before your last expected frost date. Set out transplants after all danger of frost has passed. Gardeners in frost-free climates can transplant a second crop in midsummer for a fall harvest.

Give the plant some time to branch out before transplanting. You can plant seedlings about 1 in. (2.5 cm) deeper than they were in their pots because they have small roots that grow at their base.

They fare well in a moderately rich, well-draining soil. Give them plenty of water while growing to plump up those thick walls, but do not add much fertilizer unless your soil is very poor.

How to harvest Red is the color of 'Red Mini Bell' when fully ripened. If you leave all the peppers on the plant to ripen, you will be able to harvest only a couple of times. If you enjoy the sharper flavor of green peppers, you can pick the fruits earlier and get a larger yield. The piquancy of green peppers goes wonderfully with slightly tart fillings. Cut the peppers from the plant to avoid damaging the stem.

Tip Although the plant often produces 20 or more peppers, they probably will not all appear and ripen at the same time. If you think you will need a lot of peppers at once for entertaining, plant several plants.

Others to try Lucina Cress gave two additional colors of mini bells to Seed Savers. 'Mini Chocolate Bell' peppers are a deep burgundy, almost chocolate, in color. These are less sweet and more musky than 'Red Mini Bell'. 'Mini Yellow Bell' has the sharper bite of yellow bells.

Exposure:
Full sun

Ideal soil temperature:
75–80°F (24–27°C)

Planting depth:
1–1½ in. (2.5–4 cm)

Days to germination:
10–14 days

Spacing:
18 in. (45.5 cm)

Days to maturity:
75–95 days

The perfect little bite, 'Red Mini Bell' matures quickly and keeps producing.

POTATOES AND SWEET POTATOES

At the top of many lists of comfort foods, potatoes have a deep, earthy aroma that convinces you they took full advantage of the rich garden soil. Digging potatoes is like a treasure hunt, as you poke through the fluffy soil to find every last plump tuber and every delectable baby spud. Eating freshly dug and cooked potatoes is an eye-opening experience, with complex flavors that have not had time to dissipate in storage and travel.

Because potatoes are grown from pieces of the tuber, they are not only heirlooms, but also clones. When you eat a potato, you are truly tasting Old World flavor. Although potatoes originated in South America at least 3000 years ago, they took a circuitous route to North America. The Spanish encountered them while looking for gold in Peru and eventually brought them home to Spain. Potatoes were used throughout Europe, although mostly as fodder for livestock. They are believed to have been brought to Virginia in 1621, along with tobacco, by British traders based in Bermuda. Tobacco caught on, but potatoes would have to wait until a group of Irish Presbyterian immigrants began growing and eating them in a settlement in Londonderry, New Hampshire, in 1719.

Potatoes are not related to sweet potatoes, although sweet potatoes also come from South and Central America. Christopher Columbus introduced sweet potatoes to Europe as "batatas" or "patate," which is possibly how they became confused with common potatoes. Sweet potatoes are especially popular in tropical and subtropical areas because they thrive in hot weather.

'German Butterball' potatoes prove good things really do come in small packages. These netted, oval spuds are not much larger than a flattened tennis ball, but they are worth the peeling. When cooked, the flaky yellow flesh truly looks and tastes as though it had been baked with butter.

This is a late season variety with sprawling vines. The tubers also tend to take off in all directions, so be sure to hunt for every last one.

It is difficult to believe something this buttery grows underground. 'German Butterball' spuds are not the largest of potatoes, but they are among the best.

Exposure:
Full sun

Ideal soil temperature:
45–50°F (7–10°C)

Planting depth:
2–3 in. (5–7.5 cm)

Days to germination:
7–10 days

Spacing:
8–10 in. (20–25 cm)

Days to maturity:
105–115 days

Potato 'German Butterball'
Solanum tuberosum

Flavor Rich and buttery on their own, these potatoes can be prepared as simply as you like and will always taste delicious. They have the earthy flavor of fresh potatoes with a tender, almost creamy texture. The earthy potato aroma is particularly prominent with roasting.

Growing notes Plant seed potatoes 1–2 weeks before your last expected frost date, as long as the soil has dried to a crumbly texture. Gardeners in frost-free zones can plant potatoes in early spring for summer harvest or late summer for a late fall harvest.

Potatoes like a well-draining, light, sandy soil. Although 'German Butterball' has some resistance to scab, a slightly acidic soil, pH 5.0–6.8, will offer even more scab resistance. To avoid spreading disease, use only certified seed potatoes.

Small potatoes can be planted whole. Cut larger potatoes into pieces with at least two eyes, and plant them immediately or allow them to dry for a day. Pile soil on top of potatoes as they grow. The hill of soil keeps the emerging tubers away from sunlight and creates a fluffy bed in which they can grow. Start with a trench or hill up as you go along. At planting time, cover the seed potatoes with 2–3 in. (5–7.5 cm) of soil. As the plants grow to 6–8 in. (15–20 cm), cover or hill up the bottom 3–4 in. (7.5–10 cm) with mulch or soil. Repeat the hilling two more times, at 2–3 week intervals.

Feed your potatoes when the plants first sprout and again when they start to flower. Keep them watered, but do not let the soil remain wet. Tubers start to form after the flowers bloom. At that point, cut back on watering so the potatoes do not rot or crack.

How to harvest For best storage, leave the tubers in the ground until the vines have died back. Gently loosen the soil with a fork and use your hands to feel around for the potatoes.

Tip You can still plant potatoes that have already sprouted or turned green from sun exposure. Be careful that you do not knock off any sprouts.

Others to try 'Carola' produces lots of creamy yellow potatoes that are especially tasty when they are young. 'Yellow Fin' is another great, all purpose, buttery tasting potato. 'Yukon Gold' is a classic yellow potato with a drier flesh that is great for baking.

This little fingerling is a treat for all the senses. A great all-purpose potato, its moist flesh produces a sweet, sumptuous aroma during cooking. Unlike most purple potatoes, 'Purple Peruvian' retains its purple color inside and out when cooked, adding a colorful surprise to dishes. And the color offers some powerful antioxidants.

Peruvians grow hundreds of types of potatoes. You might not have space in your garden for hundreds, but you should make room for 'Purple Peruvian', an easy growing, late-season variety.

Potato 'Purple Peruvian'

Fingerling potato
Solanum tuberosum

Flavor These potatoes bring a nutty earthiness to the table. They can be a little moist and make a good choice for frying or baking. The color cannot be beat for potato salad, and the mellow flavor is a great foil for dressings. Use a light touch and do not overcook them.

Growing notes Seed potatoes can be planted 1–2 weeks before your last expected frost date, after the soil has dried out. Gardeners in frost-free areas can plant potatoes in early fall for a winter harvest. They prefer a well-draining, light, sandy soil. Be sure to use certified seed potatoes.

Plant in a shallow trench, and hill soil over the potatoes as they grow. Start by covering the seed potatoes with 2–3 in. (5–7.5 cm)

Exposure:
Full sun

Ideal soil temperature:
45–50°F (7–10°C)

Planting depth:
2–3 in. (5–7.5 cm)

Days to germination:
7–10 days

Spacing:
6–8 in. (15–20 cm)

Days to maturity:
90–110 days

'Peruvian Purple' holds its color when cooked, unlike other blue potatoes.

of soil. As the plants grow to 6–8 in. (15–20 cm), cover or hill up the bottom 3–4 in. (7.5–10 cm) of the plants. Repeat the hilling twice more, at 2–3 week intervals. The deep soil keeps sunlight off the emerging tubers and makes a nice fluffy bed in which the tubers will grow.

Feed potatoes when the plants first sprout and when they start to flower. Keep them watered, but not so much that the soil stays wet. The tubers start to form after the flowers bloom. At that point, cut back on watering so the potatoes do not rot or crack.

How to harvest Leave the tubers in the ground until the vines have died back. Gently loosen the soil with a fork and use your hands to feel around for the potatoes.

Tip Fingerlings such as 'Purple Peruvian' are good choices for growing in containers. Put a few inches of soil at the bottom of a large container, such as a barrel or clean garbage can, and then add the seed potatoes and cover with a couple of inches of soil. Keep topping with soil as the vines grow. When the time is right, tip over the container to harvest the potatoes.

Others to try 'All Blue' is great for baking; the blue color tends to disappear with cooking. 'La Ratte' is a moist and flaky French fingerling. 'Swedish Peanut' has drier flesh with a surprising peanutlike flavor.

'Georgia Jet' is a colorful end-of-season delight. Its dusky rose skin and creamy orange flesh are as inviting as a juicy citrus fruit. But 'Georgia Jet' positively says fall.

Sweet potatoes are members of the same family as morning glories, although their beauty is mostly hidden underground rather than in their blossoms. The vines can sprawl like those of their family members, but 'Georgia Jet' is considered a semi-bush variety that can be fitted into a garden or used as a ground cover. This old standard is extremely reliable and produces a crop even in areas with short growing seasons.

Sweet potato 'Georgia Jet'
Ipomoea batatas

Flavor 'Georgia Jet' has a gentle, sweet squash flavor with an earthiness that sneaks up and surprises you. It will bake to a smooth and creamy consistency, and roasting 'Georgia Jet' actually intensifies its character. It also pairs delightfully well with raisins and spices such as curry.

Growing notes Sweet potatoes are usually grown from slips (small rooted pieces of tuber) that you can purchase (certified disease-free slips are best), or you can create your own by slicing a sweet potato in half lengthwise, placing it on a bed of damp potting mix, and covering it with a few inches of soil. Keep the soil warm and moist, and roots should develop within a week or so, followed by leaves. They are ready to transplant in about 6 weeks, when they reach 4–8 in. (10–20 cm) tall.

Plant the slips 3–4 weeks after your last anticipated frost date and after the soil has warmed and dried. You can encourage soil warming by covering it with black plastic. Sweet potatoes need a long, hot season and are generally not planted in the fall.

Where winters are short, slips can be started from existing vines in the fall. Cut 4–6 in. (10–15 cm) tips from your vines and root them in a jar of water. When the roots develop, pot them in soil and wait for planting time to transplant outdoors.

Sweet potatoes will tolerate most soil types, but a loose, sandy loam will yield the largest tubers. Do not use excessive amounts of organic matter to amend the soil, because overly rich soil can lead to fungal diseases. They also prefer a slightly acidic soil, at pH 5.8–6.0. Add some compost or a balanced fertilizer at planting time to give them a head start.

Water the plants well at planting. Do not worry if the foliage is slow to grow; the plants are establishing their root system. After the plants start to grow, keep them watered regularly, but allow the soil to dry before watering again. Too much water will cause the tubers to crack.

Few pests and problems are associated with sweet potatoes. If you have a problem with rot, consider applying a

Exposure:
Full sun

Ideal soil temperature:
70–75°F (21–24°C)

Planting depth:
Cover seedlings to a depth of about three leaf nodes

Days to germination:
Not applicable

Spacing:
10–15 in. (25–38 cm)

Days to maturity:
90–100 days

'Georgia Jet' offers a creamy rich texture when baked; tubers can grow into interesting shapes.

fungicide the next time you plant them.

How to harvest Start to harvest when the tubers are 1½–2½ in. (4–6 cm): you can loosen the surface soil and feel around to find the tubers and check their size. Harvest the entire crop when the leaves begin to yellow and die back, and be sure to dig them up before a frost.

Any tubers damaged while digging should be used immediately. Allow the rest to cure for a couple of weeks so their skins will toughen. After that, they can be stored for several months.

Tip Sweet potato leaves are edible and make a nice addition to stir-fries.

Others to try 'Frazier White' has white skin and sweet flesh and is suitable for shorter growing seasons. 'Nancy Hall' is sweet and waxy, with bright yellow flesh; it is nice baked or mashed. 'Violetta' is an extremely sweet, purple-skinned variety that bakes well.

RADISHES

Radishes are bold and bright, in brilliant shades of red and white, to rich, inky black. The perfect radish is so crisp it is almost brittle. It should have warmth, but not searing heat, with an animated, peppy flavor that pulls it all together. All that character should not be relegated to a mere salad garnish. The radish should be treated as hearty, but refreshing fare to be included in soups, sandwiches, and salads.

Radishes have a long history and were valued as a versatile root crop in ancient China, Greece, and Egypt. These were generally long, tapering black or white varieties. The round red radish put in an appearance only about 200 years ago.

It is easy to forget that radishes are a root vegetable, not just a salad ingredient or garnish. 'French Breakfast' has enough character to be treated and respected as a delectable vegetable.

Dating back to at least the 1880s, this beautiful French heirloom is an elongated type with a rosy glow that fades to an icy white. Quick and easy to grow, 'French Breakfast' stays crisp and mild more readily than round, globe radishes.

Radish 'French Breakfast'
Raphanus sativus

Exposure: Full sun to partial shade

Ideal soil temperature: 65–70°F (18–21°C)

Planting depth: 1/4 in. (0.6 cm)

Days to germination: 5–7 days

Spacing: 1–1¹/2 in. (2.5–4 cm)

Days to maturity: 20–30 days

Flavor 'French Breakfast' is more spicy than hot, with a mild, chilly pungency. It pairs remarkably well with dairy products such as butter and creamy dressings and soups. Line thin slices of 'French Breakfast' radish on a warm baguette, covered in a blanket of sweet butter with a sprinkle of salt, and you have a breakfast worthy of a sidewalk cafe.

Growing notes Begin to direct sow seeds outdoors after the ground has warmed and dried out a bit. The soil should be slightly damp, so that it forms a loose ball in your hand. Continue to succession plant every 2–3 weeks. Gardeners in warmer climates can succession plant throughout the fall and winter.

Radishes grow best in cool, damp weather. They need to grow quickly or they become woody and harsh. 'French Breakfast' radish is more forgiving than many varieties, but for the best quality, get your seeds in the ground early.

Keep the soil moist until the seed has germinated. Water them daily to keep them cool and encourage them to grow quickly. No fertilizer or amendments should be required, unless your soil is excessively poor.

Flea beetles will eat holes in radish leaves, but they will not harm the roots. Cabbage root maggots can ruin globe radishes by burrowing through them. Row covers will keep them out. It is also wise to rotate your planting areas to avoid planting in infested soil.

How to harvest The shoulders of the radishes will barely poke above the soil line when they are ready to harvest. Gently poke your finger around the top of the bulb to determine whether it has filled out. The radishes should pull out of the soil easily.

Tip Sow radish seed along with slow-growing root vegetables such as carrots and parsnips. They will mark the row with their quick growth and keep the soil from crusting over. Pulling up the radishes will help thin the incoming companion plants.

Others to try 'Champion' is a 1957 All-America Selections winner that is tender and pleasantly pungent. 'Purple Plum' is a bright purple radish with a sweet and mild flavor. 'Sparkler' is round and red with a white bottom and has mild heat; it is slow to turn pithy.

Exposure:
Full sun

Ideal soil temperature:
65-75°F (18-24°C)

Planting depth:
1–1½ in. (2.5-4 cm)

Days to germination:
5-10 days

Spacing:
18 in. (45.5 cm)

Days to maturity:
50 days

Wow your friends by growing and serving rat's tail radishes. Bite into one for waves of radish flavor in an unrecognizable form. This aerial radish does not produce an underground bulb. The part you eat is the crunchy seed pod that tastes similar to a juicy radish root.

Technically, all radish seed pods are edible, as are their flowers. But those belonging to the Caudatus group, which means "with a tail," have been bred to produce more flowers, go to seed faster, and stay tender and juicy.

This plant was brought to America from Japan in the mid-1800s, when it was called Japanese radish or Java radish, where it was thought to have originated. Seed has been cross-pollinated over the years, and today it is difficult to find a true rat's tail radish. The flavor does not vary widely with the other types, but pure seed is always valued as a true heirloom.

(Opposite) 'French Breakfast' radishes are mild flavored if planted in cool weather and grown quickly.
(Above) Rat's tail radishes produce crunchy, plump pods instead of subterranean bulbs.

UNUSUAL

Radish, rat's tail
Raphanus sativus 'Caudatus'

Flavor The pungency of rat's tail radish sneaks up on you. At first bite, you notice a green, grassy flavor, and then, Wham! You know you are eating a radish.

There is definite heat there, and a whole lot of fresh, peppery radish flavor. It is not unpleasantly potent, but it definitely wakes up your palette. The pods lose some of their potency when cooked, but they maintain their crunch and good flavor.

Growing notes Direct sow seeds in the garden in late spring. Make successive plantings every 3–4 weeks into mid-summer. This plant thrives in heat. Gardeners in frost-free climates can sow seeds in early spring and again in late summer to harvest in the fall.

Rat's tail radishes need a sunny spot and water. They are quick to germinate in early summer and can reach heights of 2–5 ft. (0.6–1.5 m). The plants are heavy and spreading, so some type of support is required or they will sprawl on the ground.

Plants are virtually pest-free. Because no bulb forms, cabbage maggots are not a problem as with typical radishes.

How to harvest Harvest the pods while they are young and tender. They are best when they grow no larger than a slim green bean. The young pods will snap off easily, but older pods should be cut off the plant. Pods will be produced for at least 2 months if you harvest regularly. Do not plant rat's tail in the vicinity of other radishes if you intend to save seed.

Tip The flowers are butterfly magnets and a great way to attract pollinators to your garden.

Others to try Seeds from other countries are offered under different names, such a 'Mougri' and 'Snake', but they are probably the same plant. 'Munchen Bier' produces an edible white tap root as well as crunchy seed pods.

'Red Meat' is a juicy, slightly sweet, and stunning globe radish. Creamy white and green, it tapers to a long, pink tap root. When sliced, it reveals a glowing interior that can range in color from pale rose to a shocking crimson, hence its nicknames—watermelon radish and beauty heart.

'Red Meat' is a Chinese variety. Unlike many of the Asian winter radishes, it is more turnip shaped than long or thick, with a refreshing sweetness rather than heat.

VERSATILE

Radish 'Red Meat'
Watermelon radish, beauty heart
Raphanus sativus (Longipinnatus Group)

Flavor Unlike most winter radishes that pack a punch, 'Red Meat' is sweet and crunchy, with just enough lingering heat to remind you that it is a radish. It is best when eaten raw and can be featured on a hors d'oeuvre plate or sliced as a serving cracker for a topping.

Growing notes 'Red Meat' needs to grow in cool weather and has a longer growing season than typical globe radishes, so it seldom does well when sown in the spring. Gardeners in cool climates can direct sow seed in

Exposure:
Full sun to partial shade

Ideal soil temperature:
55–65°F (13–18°C)

Planting depth:
1/4–1/2 in. (0.6–1 cm)

Days to germination:
9–15 days

Spacing:
1–3 in. (2.5–7.5 cm)

Days to maturity:
45–60 days

Bite into 'Red Meat' for a burst of flavor, with a refreshing, mild zest.

midsummer to late summer to harvest in the fall. Those in frost-free climates can sow seed in late summer through fall.

Plant seeds directly in the garden. Plants prefer a moderately rich soil that is loose and well-draining, but they will tolerate clay soil. Keep the soil moist until the seed has germinated, and water regularly throughout the growing season. A layer of mulch will help to keep the roots cool. No fertilizer or amendments are required unless your soil is excessively poor.

Because this radish is in the ground longer than globe radishes, it is more prone to flea beetle damage, but the more damaging cabbage root maggots should be gone by the time it is planted. It is still wise to plant this and all winter radishes in a spot where no other radishes or cabbage family members were planted that season.

How to harvest 'Red Meat' will start to push its shoulders above the soil when it is ready. The bulbs will reach about 4 in. (10 cm) in diameter, but you can start harvesting them as soon as they reach 1¹/₂–2 in. (4–5 cm).

Harvesting the smaller roots will give the remaining roots more room to fill out.

Tip You might see 'Red Meat' sold as red daikon; it can be used in recipes calling for daikon radishes, whether fresh or roasted.

Others to try 'Green Meat' is a green and white daikon-type radish. 'Helios' is sweet, tender, and pale yellow with white flesh. 'White Winter' is a blunt-tipped solid white Chinese winter radish, with a lot of bite.

The imposing inky black skin of 'Round Black Spanish' radish makes a great foil for its frosty white flesh. Still, nothing prepares you for the wallop of heat this radish can deliver, as the spicy aroma swirls around your taste buds.

This exceptionally hardy winter vegetable can be overwintered in the ground and enjoyed the following spring. Although uncommon today, 'Round Black Spanish' was once one of the most popularly grown root crops. It has been grown in North America since at least the early 19th century and is still popular in many Passover recipes.

SPICY

Radish 'Round Black Spanish'
Raphanus sativus

Flavor 'Round Black Spanish' can be used raw, sliced, or shredded. It is crisp, spicy, and bracingly hot. The skin may look woody, but it is surprisingly tender and does not need to be peeled before eating. The pungent flesh will need a little taming, however. Salting will cause slices to weep, and then draining the liquid will temper the heat. Some people simply soak the slices overnight in ice water. If you are a heat lover, you can eat slices with a sprinkling of salt. It also cooks well and adds an unexpected kick to stews and stir-fries.

Growing notes 'Round Black Spanish' is a slow grower, and if planted in the spring, it will start to bolt as the days warm up. It is usually planted in midsummer to late summer and allowed to fill out as the days shorten and nights cool, and it is harvested from fall into winter. Gardeners in frost-free climates should wait until their cool season, planting in late summer through fall.

Plant seed directly in the garden. Keep the soil moist until the seed has germinated and throughout the growing season. A layer of mulch will help keep the roots cool. It prefers a moderately rich soil that is loose and well-draining, although it will tolerate clay soil. No fertilizer or amendments are required unless your soil is excessively poor.

This radish's pungency tends

Exposure:
Full sun to partial shade

Ideal soil temperature:
55–65°F (13–18°C)

Planting depth:
1/4–1/2 in. (0.6–1 cm)

Days to germination:
9–15 days

Spacing:
4–6 in. (10–15 cm)

Days to maturity:
50–60 days

'Round Black Spanish' radishes are black on the outside, but deceptively cool and white inside.

to keep most pests at bay, although the leaves may see some flea beetle damage.

How to harvest The radishes will start to expand and push up above the soil line when they are ready to be dug. You can start harvesting when they reach 2–3 in. (5–7.5 cm) in diameter. You can also leave some to overwinter in the ground with a little mulch protection, to harvest in the spring.

Tip These radishes store well as long as they are kept dry. They will even mellow a bit the longer they are stored.

Others to try 'Hilds Blauer' is a beautiful, purple-skinned radish with a mild bite that stores well into winter. 'Green Luobo' is a long, green Daikon-shaped winter radish that is tender, with mild heat. 'Long Black Spanish' is similar, but a bit longer at 8–9 in. (20–22.5 cm), with tapered roots that are even more pungent.

'White Icicle' is a refreshing, juicy radish, with icy, cool white fingers and a mildly spicy bite. Grown since the mid-1800s, this long, slender radish can grow to 4–5 in. (10–12.5 cm) and retains its delicate balance of flavors.

'White Icicle' has gone by many picturesque names, such as 'White Italian' and 'White Naples', but its beauty is more than skin deep. No mere novelty, it grows well in less-than-perfect soil, is slow to bolt, and consistently delivers a sweet, pleasant burst of radish gusto when eaten.

Radish 'White Icicle'
'White Italian', 'White Naples'
Raphanus sativus

Flavor If you like the bite of radishes but not the assault of too much heat, 'White Icicle' is for you. Their frosty, arctic appearance is appropriate for their crisp, cool taste, with a spicy aroma and a crunchy tang. They are tender and refreshing when used fresh but can also be tossed in stir-fries or lightly braised for a side dish.

Growing notes Direct sow seed in the garden in spring, as early as possible and as soon as the ground dries to a crumbly texture. Succession plant about every 10 days until the weather turns hot. Gardeners in cool climates can start sowing again in late summer. Gardeners in warmer climates can succession plant throughout the fall and winter.

Keep the soil moist until the seed has germinated. It tolerates most soils, but a well-tilled, fluffy soil will allow it to grow long and relatively straight. Mulch the plants to keep the roots cool in summer. No fertilizer or amendments are required unless your soil is excessively poor.

Exposure:
Full sun to partial shade

Ideal soil temperature:
45–75°F (7–24°C)

Planting depth:
1/4–1/2 in. (0.6–1 cm)

Days to germination:
4–12 days

Spacing:
2–4 in. (5–10 cm)

Days to maturity:
25–35 days

'White Icicle' radishes can be so white, they are almost clear, like their namesake.

Flea beetles will eat holes in the leaves, but they will not harm the roots. Watch out for cabbage root maggots; use row covers to keep them out. It is also wise to rotate planting areas.

How to harvest Radish shoulders will start to emerge above the soil when they are ready to harvest. Do not be alarmed if the shoulders are green; they turn that color when exposed to the sun, but the taste and quality are unharmed. A variant form will turn purple on top, but it has the same flavor. If your soil is loose, you can twist and pull the radishes out of the ground; otherwise, gently loosen the soil with a fork before lifting.

Tip To keep freshly harvested radishes crisp, place them in a pail of cold water immediately after harvesting.

Others to try 'Cincinnati Market' is a mild, sliver-thin, long, red radish. 'Ostergruss Rosa' is a spicy-sweet radish that resembles a rosy carrot.

RHUBARB AND SORREL

Many fruits masquerade as vegetables in the garden, but these two vegetables do great imitations of fruits, with the tangy tartness of citrus and berries. Both are perennial plants that make themselves at home in your garden and grow larger and more delectable each year. Plant them and look forward to finding more and more ways to enjoy them as you discover how weather can affect the degree of sweetness or tartness.

Exposure:
Full sun to partial shade

Ideal soil temperature:
Below 40°F (5°C)

Planting depth:
2–3 in. (5–7.5 cm)

Spacing:
3–4 ft. (0.9–1.2 m)

When it was introduced, 'Victoria' rhubarb was heralded as the first sweet sign of spring. Imagine a winter without fresh fruit and the anticipation of seeing the first stalks of juicy rhubarb poking up in early spring. This was one of a parade of vegetables named in honor of Queen Victoria's coronation and one of the few that remain popular today.

Rhubarb 'Victoria'
Rheum rhabarbarum

Flavor 'Victoria' is a rhubarb classic. The berry-red stems are thick and succulent, with just the right balance of sweet and tart. The sweet juiciness is offset by an astringent tart quality and blends well with sweeter fruits, such as oranges and the traditional strawberry pairing, without becoming cloying. It makes a spicy addition to seasoned meat dishes, and its subtle sweetness makes it a great choice for rhubarb wine.

Growing notes Seeds and root stock can be planted in early spring, although it is rarely grown from seed. Directions provided here are for planting root stock. Gardeners in Zones 8 and warmer can have difficulty growing rhubarb, because rhubarb needs a cold chill, below 40°F (5°C), to break dormancy. You can grow it as an annual, planting it in late summer to harvest the following spring.

Given the right conditions, rhubarb is a perennial plant

'Victoria' rhubarb is a perennial plant, growing new and succulent stems each spring.

that will improve in quality and productivity as it becomes established. 'Victoria' can produce for 20 years, so find a spot where it can grow undisturbed. Prior to planting, prepare the soil with lots of organic matter to create a rich, well-draining loam. You might want to do this in the fall, so that the bed will be ready to go in early spring.

The dormant root stock looks like a gnarly clump of soil, but the plant is still alive. Cover it with soil, firm it down gently, and water it well.

Do not harvest stalks the first year. Like asparagus, the plant needs all its leaves to store energy via photosynthesis. Side dress midseason with compost or manure and again each subsequent spring.

You will need to lift and divide the plant every 3–5 years. Use a knife to divide the crown into several sections and replant.

How to harvest Starting the second year, you can harvest stalks of 'Victoria' rhubarb by pulling or cutting them off at soil level. Take only 2–3 stalks per plant at a time to allow the plant to continue to grow.

Tips The flower stalks are ornamental, but cutting them off will direct more energy back to the plant. Let it flower only if you are saving seed. Never eat rhubarb leaves, and do not eat stalks that have been hit by a frost. They contain high levels of oxalic acid, making them poisonous to humans.

Others to try 'Champagne Early' is a pale green variety popular for making wine. 'Paragon' is a low acid red variety that is easy to grow from seed. 'Strawberry' makes nice jam.

Sorrel is a perennial plant that lives for several years and will self-seed if you let it.

This tender, broad-leaved green has a tart, lemony flavor that is too often relegated to the herb garden. It looks a lot like spinach and is sometimes listed in garden catalogs as spinach dock. After you have tasted it, however, you will never confuse the two. Sorrel has a much higher oxalic acid content, which gives it a sour punch.

Sorrel has been included in American gardens from at least the early 19th century. As American tastes turned sweeter, sorrel became an occasional seasoning. But it is a green, and like many others, it is healthy and delicious.

Sorrel
Spinach dock
Rumex acetosa

Flavor Garden sorrel's flavor is undeniably tart and lemony. It adds a pleasant sourness to salads and sandwiches when used fresh. It almost melts when heated and is a classic for soups and for sauces drizzled over eggs or fish. Its acidic taste

Exposure:
Full sun to partial shade

Ideal soil temperature:
65–75°F (18–24°C)

Planting depth:
1/4–1/2 in. (0.6–1 cm)

Days to germination:
7–14 days

Spacing:
8–12 in. (20–30.5 cm)

Days to maturity:
60–90 days

makes it as easy to pair and as versatile as lemons.

Growing notes Direct sow in the garden in early spring, or start plants indoors 4–6 weeks before your last expected frost date. Gardeners in frost-free climates can also start seed in the fall.

Sorrel is a perennial plant, returning and expanding every year. Prepare the soil well before planting. Keep the soil moist until the seed has germinated and throughout the plant's growing season. Although not demanding, it does best in a rich, organic soil, so add compost in the spring or the fall before planting. In subsequent years, side dress plants in the spring with compost or composted manure.

Sorrel may bolt in warm weather and the leaves will acquire a more bitter taste. After it bolts, stop harvesting and let the plant grow and build up strength. Remove the seed heads unless you are saving seed; if left to mature, they will self-seed throughout the garden.

Lift and divide sorrel plants in the spring, slicing them into sections and replanting the divisions as needed. Sorrel plants should last for 7–10 years, but you might want to replace them every 3–5 years, because the older plants' leaves can be tough.

Leaves can be damaged by feeding larvae. Deer and rabbits also find sorrel too tempting to pass up. Use a row cover to prevent damage.

How to harvest Use sorrel as a cut-and-come-again green or slice off entire plants about 1–1½ in. (2.5–4 cm) from the base. The plants will fill back in.

Tip Sorrel can turn an odd shade of grayish-green when cooked. Add some spinach or parsley to make the color a bit more appetizing.

Others to try You will not find many named varieties of garden sorrel. 'True French' (*Rumex scutatus*) sorrel has smaller leaves with more concentrated, intense flavor, but garden sorrel makes a fine substitution. 'True French' is favored in France, but seed is seldom offered in America.

SALAD GREENS

You need never get stuck in a salad rut with so many enticing greens to toss into the bowl, with greens and reds and speckles and splotches. These are some of the fastest growing plants in your garden, ready for snipping within weeks of planting. Pluck some tender new leaves and even more will follow, promising tantalizing salads throughout the spring and summer.

Although ancient lettuce was more stem than leaves, it was still fit for Persian royalty back in 550 BC. The Romans thought lettuce was an aphrodisiac. Flavor and tenderness have improved over the years, and salad greens remain one of the most widely enjoyed vegetables in the garden.

'Apollo' arugula's sweet nuttiness is unexpected in a green often considered savory. The large, rounded leaves are softer than the usual serrated forms and have a tender-crisp texture that complements the gentle sweetness. Also called rocket or roquette and related to the radish, arugula can have an overwhelming flavor with a bitter heat. 'Apollo', however, is gently flavored enough to feature solo in a salad. This refined selection is from the masterful Dutch breeders.

Arugula 'Apollo'
Rocket, roquette
Eruca sativa

Flavor 'Apollo' gives you the peppery zip of arugula without the unpleasant muskiness of many wilder varieties. Its flavor is more nuanced and layered than typical arugulas, and it can be eaten fresh or slightly wilted atop pasta or a pizza.

Growing notes Direct sow the tiny seeds in early spring, 1–2

Exposure:
Full sun to partial shade

Ideal soil temperature:
50–75°F (10–24°C)

Planting depth:
1/8–1/4 in. (0.3–0.6 cm)

Days to germination:
10–12 days

Spacing:
Broadcast somewhat evenly

Days to maturity:
40–45 days

'Apollo' arugula can withstand some summer heat, especially if it is watered well.

weeks before your last expected frost date, or start them indoors about 4–6 weeks before transplanting out. Succession planting every couple of weeks will extend the harvest. In Zones 8 and warmer, fall is the best time to start arugula seeds. Gardeners in cooler climates can grow a fall crop by starting plants in pots in late summer in a shady location.

Arugula grows quickly, and you do not need to thin the plants. Because it is grown for its leaves, it requires a rich soil and lots of nitrogen to stay fresh and green. Plenty of organic matter worked into the soil should be enough of a boost to get arugula through its short growing season. If your plants start looking pale, water them with some compost tea.

In addition to supplying water, timing is the essential ingredient for growing great arugula. Seeds need to be set out late enough so the young plants are not zapped by frost, but early enough so they mature before they bolt in the heat. Unfortunately, the weather does not always cooperate.

'Apollo' grows in decorative rosettes. The leaves stay flat and do not grow tall, but the flower stalks will shoot up 12 in. (30.5 cm) or more. The airy, white flowers will signal the end of the growing cycle, but the blossoms are edible and tangy, too.

How to harvest Snip outer leaves as you need them. Arugula is a cut-and-come-again plant, so you should be able to get 4–5 cuttings per plant if you do not let the plants go to seed. Hot weather can cause the plants to bolt; keep them well watered and harvest often to prolong production.

Tip If you are having trouble keeping arugula from bolting in a sunny garden, plant them between taller vegetables, such as broccoli or peppers, which can help shade the plants. By the time these vegetables spread out, you will have enjoyed several weeks of arugula harvests.

Others to try Arugula seeds without a particular variety name will give you the traditional sharp arugula you may prefer. 'Olive Leaf' produces smooth, pale green leaves that are tender with a pleasant bite. 'Sylvetta' is a wild variety with short, rough leaves and is extremely slow to bolt.

It takes longer to say 'Catalogna Puntarelle Stretta' than to grow it. This strappy chicory falls somewhere between the refinement of endive and the wild abandon of dandelions. It does not have the jagged, bushy leaves of frisée (curly endive) and does not require blanching to make it palatable. A springtime treat in Italy, it is often prepared with anchovies, but you can enjoy it throughout the summer, with or without the fish.

You may have seen wild chicory growing along the side of the road. It is a tall, gangly plant with a pretty blue flower. Chicories will readily cross-pollinate and self-seed.

Chicory 'Catalogna Puntarelle Stretta'
Cichorium intybus

Flavor The smooth, elongated leaves of this selection start out sweet and develop a pleasant bitterness, not the astringent sourness often associated with chicory. It is a tangy addition to salads, but it also makes a wonderful side dish when wilted and seasoned. The green part of the leaves are often stripped off, with just the crunchy, white ribs sent to the table.

Growing notes Start seed indoors, 6–8 weeks before your last expected frost date. Or direct sow seeds in the garden in early spring and succession plant every couple of weeks into early summer and again in early fall, for continued harvests in cooler areas. It can also overwinter and be ready to harvest in the spring. Gardeners in Zones 8 and warmer should plant in late summer and succession plant through fall.

Lightly cover the seed and keep the soil moist until the plants have sprouted. You can seed thickly and thin seedlings when they have reached 2–4 in. (5–10 cm) tall.

The flavor and texture of chicory is best if it is grown quickly. The leaves can get tough and chewy when growth slows in the summer heat. It prefers cool days, a rich organic soil, and lots of water and sunshine. It will not need feeding if planted in rich soil. If necessary, feed with a high nitrogen fertilizer or compost tea.

How to harvest The best flavor develops when the heads are allowed to mature. You can cut outer leaves as needed, but it will develop its full array of complex flavors if it is allowed to form a loose head during the cooler days of fall.

Tip If chicory is too bitter, try soaking it in cold water for an hour. The chill will lessen the bitterness, as will waiting to harvest in the fall.

Others to try 'Catalogna Frastagliate' is slightly more bitter, with a serrated leaf. 'Catalogna Galatina' is called the pine cone chicory for its thick, white ribs and spiraled head.

Exposure:
Full sun

Ideal soil temperature:
55–75°F (13–24°C)

Planting depth:
1/8–1/4 in. (0.3–0.6 cm)

Days to germination:
2–15 days

Spacing:
8–12 in. (20–30.5 cm)

Days to maturity:
50–80 days

The strappy leaves of 'Catalogna Puntarelle Stretta' turn golden in the sunlight.

If the idea of eating deer tongue does not tempt you, call this lettuce by its alternate name, 'Matchless'. This is a loose leaf, bibb type lettuce, which should tell you that it is tender. The reference to a deer's tongue comes from the leaf's elongated triangular shape. A thick midrib running down the leaf adds crunch, and a slight ruffle edges the leaf.

Before you dismiss lettuce as simply filler, keep in mind that Persian kings of 550 BC feasted on lettuce. (Well, perhaps feast is too strong a word, but they certainly served it at feasts.) Lettuce is an ancient crop, although early lettuces had tall, thick stems and smaller leaves, much like Chinese stem lettuce. Homegrown varieties remind us that there is more to lettuce than a bag of prewashed greens.

Exposure:
Full sun to partial shade

Ideal soil temperature:
65–75°F (18–24°C)

Planting depth:
1/8–1/4 in. (0.3–0.6 cm)

Days to germination:
6–14 days

Spacing:
Cut-and-come-again lettuce seed can be broadcast; if grown as a head, thin to 4–6 in. (10–15 cm)

Days to maturity:
45–75 days

The leaves of 'Deer Tongue' lettuce are tender with an elongated shape.

Lettuce 'Deer Tongue'
Bibb lettuce, 'Matchless'
Lactuca sativa

Flavor 'Deer Tongue' grows in a succulent little rosette. The delectable green leaves offer both the tender, buttery quality of a bibb lettuce and the intense green flavor of a sturdier romaine. Use 'Deer Tongue' to dress up a salad or a sandwich.

Growing notes Direct sow as soon as the soil is dry enough to make furrows. Make successive plantings every 10–14 days. Start seeding again in late summer for a fall harvest. Gardeners in Zones 8 and warmer will have better luck growing lettuce from late summer into winter. 'Deer Tongue' is not a quick bolter. It is tolerant of heat and even handles a little drought without going bitter. Although it is more heat tolerant than many lettuce varieties, it will eventually go to seed once the weather turns hot. To grow it through the summer, plant it in the shade of taller plants and keep it well watered.

Barely cover the seeds with soil. Lettuce germinates faster when exposed to some light, but a little soil helps keep the seeds moist. Lettuce is shallow rooted and, unlike most plants that like to be watered deeply, it grows best with frequent, shallow watering.

Because lettuce is grown for its leaves, apply a high nitrogen fertilizer such as fish emulsion or rich compost.

How to harvest 'Deer Tongue' is prolific, so you can cut as many leaves as you want and they will keep coming. Remove the outer leaves and the center will continue to grow. If you are after a full head, cut the rosette at the soil line.

Tip The best time to harvest lettuce is first thing in the morning, when it is full of moisture. If you must pick it later in the day, it helps to water the lettuce about 30 minutes before harvesting.

Others to try 'Mascara' is sweet, with brilliant, burgundy oak-shaped leaves. 'Oak Leaf' produces delicate, pale green, oak-shaped leaves. 'Red Deer Tongue' has a crunchy, sweet flavor, with a reddish tint.

'Marvel of Four Seasons', or 'Merveille des Quatre Saisons', is a gorgeous example of the understated French heirloom. The ruffled leaves are a deep burgundy color at the tips. You get the best of all worlds in a buttery loose-leaf lettuce with a succulent, crisp center. Lettuce is such an easy vegetable to grow, that it seems silly to buy those bags of limp greens in the grocery store. It is one of the few vegetables you eat without cooking, so freshness really matters.

Dating back to the mid-1800s, this lettuce gained its following by growing well in all kinds of weather, except freezing temperatures. It keeps sending out tender, sweet leaves as long as you keep cutting it. It also matures into a head that could rival any cabbage rose in beauty.

Lettuce 'Marvel of Four Seasons'

'Merveille des Quatre Saisons'
Lactuca sativa

Flavor The texture of lettuce often affects how we perceive its taste, and 'Marvel of Four Seasons' has a nice balance of crispness, buttery texture, and juicy moisture. This butterhead variety's leaves are almost silky smooth, with a delicate fresh flavor. It is not bad looking, either.

Growing notes Direct sow seeds as soon as the soil is dry enough to make furrows. Make successive plantings every 10–14 days. Lettuce goes to seed once the weather turns hot. To grow lettuce in the summer, plant it in the shade of taller plants and keep it well watered. Start seeding again in late summer for a fall harvest. Gardeners in Zones 8 and warmer will have better luck planting lettuce from midsummer into winter.

Barely cover the seeds with soil. Lettuce germinates faster when exposed to some light, but a little soil helps keep the seeds moist. Lettuce is shallow rooted and, unlike most plants that like to be watered deeply, it grows best with frequent shallow watering.

Since lettuce is grown for its leaves, feed with a high nitrogen fertilizer such as fish emulsion or rich compost.

How to harvest Snip leaves when they are large enough to suit your taste. Remove the outer leaves and the center greens will continue to grow. It makes a beautiful loose head if allowed to grow.

Tip 'Marvel of Four Seasons' will develop more color in cool weather. It will still have excellent flavor in summer's heat, but it will not be as showy.

Others to try 'Bronze Arrowhead' has tender, narrow, lobed leaves. 'Pablo' has wide, ruffled leaves and is slow to bolt. 'Red Salad Bowl' has crisp, lobed, bronze leaves and is slow to bolt.

Exposure:
Full sun to partial shade

Ideal soil temperature:
65–75°F (18–24°C)

Planting depth:
1/8–1/4 in. (0.3–0.6 cm)

Days to germination:
6–14 days

Spacing:
Cut-and-come-again lettuce seed can be broadcast; if grown as a head, thin to 4–6 in. (10–15 cm)

Days to maturity:
55 days

Although the color of 'Marvel of Four Seasons' can fade in the sunlight and heat of summer, the crisp taste abides.

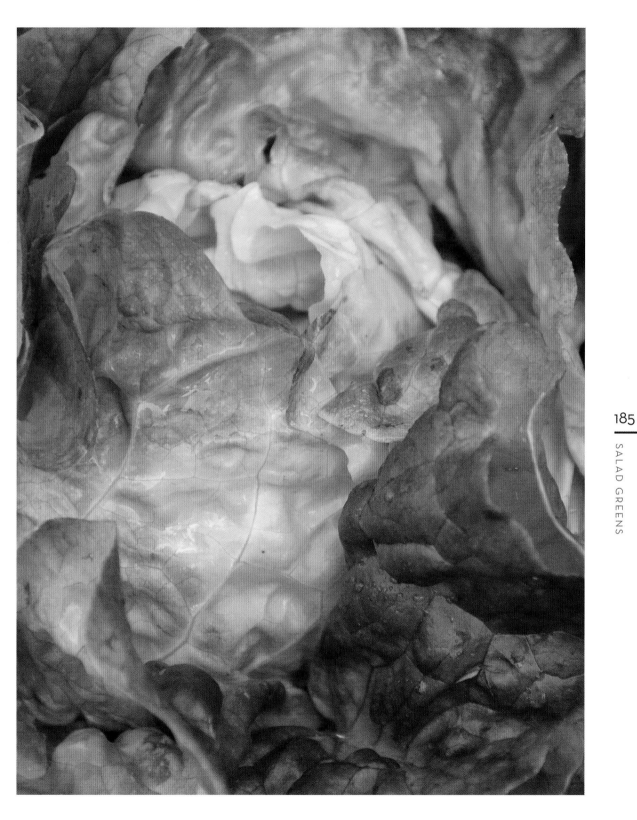

'Bistro' mache perks up with the slightest hint of spring-time sun. The fresh, herbal character of the delicate young leaves will remind you why you grow vegetables. Plant a lot of seeds, because unlike most mache varieties, 'Bistro' laughs off heat and stays sweet and tender well into summer.

Mache, or corn salad, has been cultivated since the Stone Age. How's that for an heirloom? It has been grown in the United States only since the 1700s, but other cultures have embraced it as one of the earliest greens to grow. It was named corn salad for its habit of growing freely in the fields.

The wild form of mache is quite good, but that has not prevented us from cultivating more refined forms to suit our needs. Several varieties exist, but seed can be difficult to find. Mache is generally grouped into two camps: large-seeded and small-seeded. The small-seeded varieties require cool weather, and perhaps because nothing else is growing when mache makes its first appearance, cool weather varieties have been favored in home gardens. But my favorite is 'Bistro', a large-seeded, elongated leaf variety that stays tender and tasty well into summer.

Mache 'Bistro'

Corn salad, lamb's lettuce
Valerianella locusta

Flavor The flavor of 'Bistro' is reminiscent of peanuts. The texture is tender, delicate, and buttery when fresh, so if you opt for cooking, simply wilt the leaves. Fresh, crisp 'Bistro' wakes up a salad with garden fresh greens when little else in the garden is awake.

Growing notes In spring, you can direct sow seeds as early as the ground can be worked and continue to succession plant into midsummer. Most mache varieties are sown in early fall, allowed to overwinter, and harvested the following spring. The beauty of 'Bistro' is you can do that, but you can also sow in the spring and into early summer for a pro-longed harvest. Consider mache a succulent weed: broadcast the seed, give it some water and sun, and it grows beautifully.

How to harvest Snip leaves whenever they are large enough to suit your taste. Remove the outer leaves and the center will continue to grow. When the weather heats up, pull up the whole plant to make room for something else.

Tip With a little protection, you can harvest mache all winter long.

Others to try 'Piedmont' is a similar large-seeded variety with good heat tolerance. 'Verte de Cambrai' readily self-sows for years of harvesting. 'Verte d'Etampes' is a thick-leaved, small-seeded variety that lasts longer than most.

Exposure:
Full sun to partial shade

Ideal soil temperature:
50–75°F (10–24°C)

Planting depth:
1/8–1/4 in. (0.3–0.6 cm)

Days to germination:
10–12 days

Spacing:
Broadcast somewhat evenly

Days to maturity:
45–50 days

Tender, pale green 'Bistro' mache leaves are one of the earliest vegetables to appear in the spring garden.

Mizuna's loose, serrated, bright green leaves promise a cool treat in the summer garden. Likely Chinese in origin, mizuna, also known as kyona, is a mustard that is usually associated with the Kyoto region of Japan, where it is grown for salads, soups, pickled greens, and other dishes. Although all varieties have serrated leaves, 'Early' is downright jagged, growing in frilly rosettes, like a well-behaved dandelion. It has become a darling of farmers' markets because it is easy to grow and attractive.

The tender leaves of 'Early' mizuna are often tossed in with lettuce mixes to dress them up and add nutritional value. In Japan, it is often pickled in salt brine or cooked, especially in soups. Despite its name, 'Early' mizuna is heat tolerant and holds well throughout the summer.

Exposure:
Full sun

Ideal soil temperature:
45–85°F (7–30°C)

Planting depth:
1/4–1/2 in. (0.6–1 cm)

Days to germination:
4–7 days

Spacing:
4–6 in. (10–15 cm)

Days to maturity:
30–40 days

'Early' mizuna grows in a nice, frizzy clump and does well in heat and humidity.

Mizuna 'Early'

Japanese greens, kyona
Brassica rapa subsp. *nipposinica* var. *laciniata*

Flavor This mizuna is delicate, crisp, and piquant. Its spicy mustard bite is offset by a juicy coolness, similar to a gentle escarole, without the effort of blanching. It is great fresh and adds some zip to cooked dishes. Add it to dishes toward the end of cooking to avoid overcooking.

Growing notes Direct sow seed in the garden beginning in early spring, a few weeks before your last expected frost date. You can succession plant every 2 weeks. Plants can be started indoors, 2–4 weeks before the last frost date, but they grow just as quickly by direct seeding. Gardeners in frost-free climates can continue growing 'Early' mizuna throughout the fall and mild winters.

This is one of the easiest mustards to grow. Because the plants are loose bunches of leaves, they need a rich, fertile soil and lots of water. The plants grow quickly and will not need supplemental fertilizer unless you allow them to regrow after cutting. In that case, a high nitrogen fertilizer applied every 2–4 weeks will keep them lush.

Few pests bother the zigzagged leaves. The plants are susceptible to the usual brassica family diseases, especially club root, so rotate your cabbage family plants to prevent a buildup of the organisms in the soil.

How to harvest You can harvest individual leaves as needed, or cut the whole rosette at ground level. 'Early' mizuna will regrow several times before eventually going to seed.

Tip A great container plant, its rosettes fill out nicely, and the shallow roots can easily handle the confined space.

Others to try Mizuna seed can be difficult to find, and many are the same or similar varieties sold under different names. The main difference is between thin, delicately flavored leaves such as those of 'Early' and the slightly wider leaves, such as those of 'Kyona', with a stronger peppery flavor.

Radicchio is a savory green that adds sharpness and substance to salads and side dishes. 'Rossa di Treviso Precoce' intensifies in flavor and sweetens its bite as it reddens. *Precoce* means early in Italian, and this radicchio does not keep you waiting. No tedious tying and blanching are necessary to coax it into a tantalizing piquancy.

Radicchio was popular with the ancient Egyptians and Romans. Technically, radicchio can be called Treviso only if it is grown in the Treviso area of Italy, but 'Rossa Backyard Precoce' will provide plenty of zest for your table regardless of its origins.

Radicchio 'Rossa di Treviso Precoce'
Cichorium intybus

Flavor The mild bitterness of this radicchio is best highlighted against salty and creamy foods such as cheeses and risotto, and cured hams such as prosciutto. Grilling brings out the natural sugars and softens the bite.

Growing notes Direct sow in the garden in early spring. Early plants will provide a succession of green leaves. Sowing seed in late autumn about 4–6 weeks before the first expected frost date will produce the sweetest red leaves, but with a shorter season. Gardeners in Zones 8 and warmer will have better tasting radicchio if it is planted fall through winter.

Lightly cover the seed and keep the soil moist until the plants have sprouted. You can seed thickly and thin seedlings when they have reached 2–4 in. (5–10 cm). This radicchio grows sweetest in cool days and thrives in rich organic soil with lots of water and sunshine. Leaves stay flavorful throughout the summer if you keep the plants well watered.

'Rossa di Treviso Precoce' grows quickly and requires no feeding in rich soil. If necessary, add a high nitrogen fertilizer or compost tea.

How to harvest 'Rossa di Treviso Precoce' will eventually form a heart, or head, but you need not wait; you can start harvesting the outer leaves after they reach 4–5 in. (10–12.5 cm) tall and continue harvesting throughout the summer. The leaves stay green until the temperatures start to dip. Red mottling slowly gives way to intense burgundy colors as the leaves are sweetened by the cool air. You can harvest the whole heads or leave some in the garden to return and harvest the next year. Watch for slugs in moist soil.

Tip Spacing plants closely together, 3–4 in. (7.5–10 cm) apart, will encourage them to begin to form heads and will help keep the inner leaves tender.

Others to try 'Castelfranco' forms a speckled, open-head rosette with crispy leaves. 'Chioggia' produces mild, round heads that are great for eating fresh. 'Rossa di Verona' has mild radicchio flavor in a cut-and-come-again variety.

Exposure:
Full sun

Ideal soil temperature:
45–75°F (7–24°C)

Planting depth:
1/8–1/4 in. (0.3–0.6 cm)

Days to germination:
2–15 days

Spacing:
8–12 in. (20–30.5 cm)

Days to maturity:
50–80 days

'Rossa di Treviso Precoce' does not start to form a head until the cool weather of fall.

Crunchy yet tender, delicate yet robust, 'Bloomsdale Long Standing' is everything a classic spinach should be. That is why, although it was introduced in the early 20th century, it is still one of the most widely grown spinach varieties.

'Bloomsdale Long Standing' has many good qualities. Most spinach varieties bolt to seed the minute the temperature starts to climb in the spring. This one grows quickly in cool weather and displays unusual heat tolerance for a spinach, so it is ready to start harvesting long before the weather warms and keeps going despite the heat. It is also easy to harvest, because the leaves stand up tall and stay cleaner than varieties that open low to the ground.

Although spinach is an ancient crop, it was first brought to America by Europeans in the early 1800s. 'Bloomsdale' spinach was named after the farm in Pennsylvania where it was developed. Its heat tolerance was improved by breeders in the Netherlands, and it was reintroduced in 1925 as 'Long Standing Bloomsdale', one of the best heat-resistant varieties that still has few rivals.

CLASSIC

Spinach 'Bloomsdale Long Standing'
Spinacia oleracea

Flavor Glossy, crinkled, dark green leaves look as cool and crisp as the weather in a spring garden. They are delicate, yet firm when fresh. This spinach has a bit of acidic sharpness that complements salty foods such as crumbly cheese and sausage. The thick leaves hold up well when cooked but are tender enough for eating fresh.

Growing notes Plant in early spring, as soon as the ground can be worked. Spinach can also be started indoors about 3 weeks before transplanting outside. You can also sow spinach in the late summer for a fall crop if you choose a shady spot to keep the spinach cool until fall. Plant seeds in partial shade or behind taller plants that will cast shade and keep the soil cool. It is slow to bolt, and extra water will keep it going a little longer. Gardeners with mild winters can seed throughout the fall.

This spinach can handle a touch of frost or even a light snow and will resume growing when the temperature creeps back up above freezing. In cold climates, it can be grown throughout the winter in a protected cold frame or even in a sunny spot indoors.

Spinach can be eaten by slugs, cabbage worms and loopers, and leaf miners. If leaf miners' trails start showing up in spinach leaves, removethe affected leaves as soon as possible.

Exposure:
Full sun to partial shade

Ideal soil temperature:
65–70°F (18–21°C)

Planting depth:
1/2–1 in. (1–2.5 cm)

Spacing:
4–6 in. (10–15 cm)

Days to germination:
10–14 days

Days to maturity:
48–50 days

Being early, delicious, and long lasting has kept 'Bloomsdale Long Standing' spinach popular for generations.

How to harvest Harvest spinach as you would a cutting lettuce. Pick a few outside leaves or cut the whole plant about 2 in. (5 cm) above the soil, and new leaves will emerge.

Others to try 'Giant Noble' is a smooth-leaved spinach that is slow to bolt.

SUMMER SQUASH AND ZUCCHINI

Zucchini and other summer squash are close relatives of cucumbers and melons. Like them, summer squash taste best when they are eaten immature and fresh. Every delicious part of the zucchini fruit is edible, from the blossom to the seed. You can enjoy them young and tender, raw, or prepared in any number of ways, from deep frying to stuffing.

Zucchini is native to Central America, where it has been grown for thousands of years. Christopher Columbus is believed to have brought seed to Europe, and the Italians bred the type of zucchini we recognize today. The word *zucchini* comes from the Italian *zucchino*, meaning small squash. The word *squash* is from the Native American *skutasquash*, which referred to something that was eaten raw. The French, who favor the tiniest of zucchini fruits, named them *courgettes*.

Exposure:
Full sun

Ideal soil temperature:
75–85°F (24–30°C)

Planting depth:
1/2–1 in. (1–2.5 cm)

Days to germination:
7–14 days

Spacing:
2–4 ft. (0.6–1.2 m)

Days to maturity:
50–55 days

Like sunny little flying saucers, this pattypan squash tastes almost as buttery as it looks. All dressed up in its scalloped skirt, 'Yellow Bush Scallop' is as tender as a zucchini, with a richer flesh and more concentrated flavor.

This old Native American summer squash was first introduced in catalogs in 1860. The plants grow wide and full, and the golden yellow fruits hide under the leaves, staying tender and protected.

Most of the colloquial names for scallop squash have something to do with baking pans, such as pattypan or pâtisson, a Provençal word for a cake baked in a scalloped mold. Newer varieties have been given splashier names, such as 'Sunburst', but they have not improved on the flavor of the prosaically named 'Yellow Bush Scallop'.

Squash 'Yellow Bush Scallop'

Pattypan squash, scallop squash
Cucurbita pepo

Harvest 'Yellow Bush Scallop' while they are small, and you can eat them whole.

Flavor This squash has a strong, nutty, herbal flavor and a tender texture that holds up well when cooked. With no need to peel and lose the beautiful golden edges, grilled slices will resemble flower blossoms.

Growing notes Direct sow seeds 2–3 weeks after the last frost and when the soil is warm. You can start seed indoors 3–4 weeks before transplanting out, but 'Yellow Bush Scallop' is fast growing and prolific and does not like to be transplanted.

Gardeners in frost-free climates can plant in early spring and also plant a late summer crop to mature in the fall and early winter.

This squash grows in a compact bush of about 2–4 ft. (0.6–1.2 m) tall. No trellis is needed, but give the plants room for air circulation. Plant the seed in hills, with three or four seeds per hill and 3–4 ft. (0.9–1.2 m) between hills.

Lots of flowers ensure good pollination. Both male and female blossoms appear on each plant, but only the pollinated female blossoms will become zucchini.

'Yellow Bush Scallop' shows good disease resistance but can be attacked by cucumber beetles and squash vine borers; consider adding a row cover early in the season.

How to harvest Harvest the fruits when they are tiny, with the flowers still attached, or wait until they get larger and scoop out the center to stuff. 'Yellow Bush Scallop' is most tender when it is less than 3 in. (7.5 cm) in diameter. They contain less moisture than zucchini and will get firmer as they age.

Tip This is one of the few squash varieties that grows small enough to plant and produces prolifically in a container.

Others to try 'Benning's Green Tint' is tender and a reliable producer. 'Golden Custard' is a similar variety with smooth, creamy flesh. 'Pattisson Panache' is a larger, later green and white scallop squash that bakes well.

Imagine a zucchini that is all dense, rich flesh and no seeds. 'Rampicante Tromboncino' is almost frightening to behold, even if you are accustomed to zucchini growing to the size of baseball bats. These squashes can reach lengths of more than 2 ft. (61 cm) and still be tasty. 'Rampicante Tromboncino', which loosely translates to climbing trombone, is a vigorous vine with long, curving fruits of about 3 in. (7.5 cm) in diameter, with a small, bulbous blossom end. What is so glorious about this squash, in addition to its size, is that all the seeds are contained in the bulb end. The long neck is nothing but tender flesh.

Exposure:
Full sun

Ideal soil temperature:
70–75°F (21–24°C)

Planting depth:
1 in. (2.5 cm)

Days to germination:
8–10 days

Spacing: 2–3 ft. (0.6–0.9 m)

Days to maturity:
60–100 days

UNUSUAL

Zucchini 'Rampicante Tromboncino'

Italian trombone squash
Cucurbita pepo

Flavor 'Rampicante Tromboncino' is no mere novelty. It has a wonderfully rich and nutty flavor with a texture that is dense, firm, and much less watery than traditional zucchini. It will also act as a sponge for stronger flavors, and it does not fall apart in cooking.

Growing notes Direct sow seeds outdoors 2–3 weeks after the last frost. Wait until the

'Rampicante Tromboncino' zucchinis can grow to the size of baseball bats and are still delicious.

soil warms to about 65–75°F (18–24°C) to sow seed outdoors. You can start seed indoors, 3–4 weeks before the transplant date, but this zucchini grows so quickly that starting indoors is not necessary. Gardeners in Zones 8 and warmer can make a first planting in early spring and a second toward the middle or end of summer.

Zucchini plants are easy growers and do not require much more than a sturdy trellis. Prepare yourself for vines in excess of 20 ft. (6 m). It takes a whole lot of vine to support a whole lot of zucchini. Covering the young plants will help keep pests such as cucumber beetles at bay, but uncover them when the flow-ers appear so the insects can pollinate them.

Only the female blossoms will turn into zucchini. The males start blooming first, so do not be alarmed if it takes awhile for the tiny trombones to start forming.

How to harvest 'Rampicante Tromboncino' zucchini can be harvested at any size, but they are most tender when the necks are about 2 in. (5 cm) wide. Even giant, curving fruits will be delicious.

Tip Squash grown hanging on a trellis will not curve as much as those grown on the ground.

Others to try Nothing is comparable to 'Rampicante Tromboncino', but fluted 'Costata Romanesco' is unusual and delicious and is a favorite of food writer and chef Deborah Madison.

198

Everyone's first reaction to 'Ronde de Nice' zucchini is "Oh, how cute." It may be dainty, but the plants are vigorous, and the round fruits grow blemish-free and produce for weeks, rewarding you with dozens of sweet and tender zucchini. This delicate and charming summer squash is the ideal solution for gardeners who love zucchini but do not want their garden (or kitchen) overwhelmed by it. This 19th century French heirloom produces round zucchini that can be harvested in sizes perfect for sautéing or stuffing.

VERSATILE

Zucchini 'Ronde de Nice'

Round French zucchini
Cucurbita pepo

Flavor 'Ronde de Nice' is sweet and nutty, with a skin almost as delicate as its flesh. Harvested while tiny, it makes a tender mouthful, with a creamy, rich flesh. The flavor intensifies as the orbs fill out to 3–4 in. (7.5–10 cm), when they can be scooped out and stuffed. Or simply slice and grill.

Growing notes Direct sow seeds in the garden 2–3 weeks after the last frost and after the

Exposure:
Full sun

Ideal soil temperature:
75–85°F (24–30°C)

Planting depth:
1/2–1 in. (1–2.5 cm)

Days to germination:
7–14 days

Spacing:
3–4 ft. (0.9–1.2 m)

Days to maturity:
45–65 days

'Ronde de Nice' squash likes to stay tucked under leaves, so keep an eye out and do not let them grow too large.

soil warms. You can also start seed indoors, 3–4 weeks before the transplant date, but 'Ronde de Nice' matures so quickly, you do not need to start plants indoors. Sow a second planting in midsummer for a later harvest. Gardeners in frost-free climates can plant earlier in the spring and also sow a late summer crop to mature in the fall and early winter.

This compact vine will climb if you give it a trellis, which should be installed when you sow the seeds. Plant the seeds in hills, with 3–4 plants per hill and 3–4 ft. (0.9–1.2 m) between hills. As with all zucchini, male and female blossoms appear on each plant, but only the female blossoms will become zucchini. The males start blooming first, so be patient if it takes awhile for squash to form.

Cucumber beetles and squash vine borers do not usually bother 'Ronde de Nice'. Covering the young plants will help keep pests at bay, but uncover the plants when the flowers appear so insects can pollinate them. In humid summers, powdery mildew can quickly stress and kill the vines.

How to harvest Harvest whenever the squash reaches the size you want. Keep harvesting and the vines will keep producing. Older fruits develop a tough skin and lose their delicate flavor.

Tip Cutting a slice off the bottom of the squash will allow it to sit flat for cooking or serving.

Others to try 'Table Dainty' is a tender 4–6 in. (10–15 cm) zucchini, striped in green, yellow, and white. 'Tatume' is a round Mexican heirloom that is delicious, disease resistant, and prolific. 'Tondo Scuro di Piacenza' is similar to 'Ronde de Nice' in flavor and a little larger.

TOMATOES, TOMATILLOS, AND GROUND CHERRIES

No other vegetable has done more to highlight heirlooms than the tomato. Unlike modern hybrids that are bred with thick skins to resist bruising in transit, tender, open-pollinated heirlooms lure you with their rich tomato scent and flavor. A sun-kissed tomato has been known to seduce more than one tomato lover into becoming a gardener.

Growing tomatoes is an addictive, competitive sport. You cannot grow just one. As with fresh melons, tomatoes' sweet fragrance tells you they are ready to harvest. Just brushing up against the plant will leave their lingering scent with you.

Tomatillos are like sassy tomatoes that make their presence known no matter how they are prepared. You can mix them with any Latin spices and instantly create a flavorful meal. With their citrusy tomato flavor, their refreshing acidity can wake up any dish and get the party started in the kitchen.

Ground cherries may resemble tomatillos, but they are much fruitier and sweetly refreshing. Like tomatillos, the fruits are protected in a papery husk and are sometimes called husk tomatoes.

'Brandywine' gets all the press, but most tomato gardeners I know are rhapsodic about 'Cherokee Purple'. The dusky colored fruits are a beefsteak size with small seeds, and they are usually a nicely uniform, round shape. They grow well in most climates. The shoulders of the tomato have a tendency to stay green, which is true of several heirloom varieties, but this does not affect their unique, delectable richness. The vines are indeterminate, though not particularly tall, and produce baseball-sized tomatoes.

'Cherokee Purple' is believed to have originated with the Cherokee people and has been known and grown since at least 1890. Unfortunately, the romance of heirloom vegetables often makes the validity of their stories questionable.

Tomato 'Cherokee Purple'
Lycopersicon lycopersicum

Flavor 'Cherokee Purple' has a smoky-sweet flavor. These dusky rose tomatoes, sometimes described as black in color, have a robust flavor not shared by fruitier, pale tomatoes. Something about the synergy of flavors in the skin and flesh gives 'Cherokee Purple' a mature fullness. This tomato is best eaten fresh. Bite through the sweetness of the skin and flesh to enjoy the tang of the juicy pulp.

Growing notes Start seed indoors 6–8 weeks before the expected transplant date. Move seedlings into individual 3–4 in. (7.5–10 cm) pots when true leaves appear. Transplant seedlings outdoors 2–4 weeks after your last expected frost. Wait until temperatures are reliably above 50°F (10°C) before setting out. Gardeners in areas with hot summers and mild winters will fare better starting tomatoes in late summer, to grow throughout the fall and into winter.

When transplanting the plants outside, bury the stems to the top set of leaves. The underground portion of the stem will send out roots and create a stronger plant.

'Cherokee Purple' fruits are heavy and the plants are fairly tall. Stake the plants when you plant them and secure the branches as they grow. It is susceptible to leaf and fruit diseases. Keep the leaves dry and do not plant them in the same location in consecutive seasons.

How to harvest Harvest fruits when they begin to feel soft to the touch and their smoky color has developed. You should be able to smell their sweetness. Although the shoulders do not turn purple, the green will deepen in color when they are ripe. Fruits are prone to cracking when the plant does not receive regular water, but cracking does not affect their taste. Water regularly to keep them looking beautiful.

Others to try 'Black from Tula' is a large Russian heirloom with dark skin and a similar

Exposure:
Full sun

Ideal soil temperature:
75–85°F (24–30°C)

Planting depth:
1/4–1/2 in. (0.6–1 cm)

Spacing:
3–4 ft. (0.9–1.2 m)

Days to germination:
6–14 days

Days to maturity:
75–85 days

'Cherokee Purple' does not turn completely purple, but its excellent smoky flavor is consistent.

smoky flavor. 'Cherokee Chocolate' is a close relative that is a little more tart, with dark shoulders. 'Cherokee Green' is a cross that stays dusky green, with a tangy flavor.

'Green Grape' tomatoes certainly resemble grapes, and they are sweet and refreshing when plucked from the vine and popped in your mouth. They have even been described as resembling large Muscat grapes. The 1 in. (2.5 cm) fruits stay green at the shoulder, hinting at ripeness when yellow veining appears at the midsection with a yellow glow at the blossom end. 'Green Grape' grows in clusters of 4 to 12 fruits.

This young heirloom was bred by Tater Mater Seeds in California and introduced in 1986. It does not fit the definition of an heirloom (at least 50 years old, with a unique story), but because it is a cross between two older heirlooms, 'Yellow Pear' and 'Evergreen', it is allowed some slack. Chef Alice Waters helped popularize 'Green Grape' by featuring it on the menu at her gourmet restaurant Chez Panisse in Berkeley, California.

Tomato 'Green Grape'

Cherry tomato
Lycopersicon esculentum

Flavor These zesty grapes have a clean acidity that makes them refreshing to eat. The color is a bit dull and muted on the outside, but slice one open and you will see its glossy glory. These are wonderful additions to a tossed salad, if you can resist eating them all as you walk from the garden to the kitchen.

Growing notes Start seed indoors 6–8 weeks before your expected transplant date. Move seedlings into individual 3–4 in. (7.5–10 cm) pots when true leaves appear. Transplant seedlings outdoors 2–4 weeks after your last frost. Wait until temperatures are reliably above 50°F (10°C) before transplanting out.

Tomatoes will drop their blossoms and will never fruit when temperatures stay above 85°F (30°C). Gardeners in areas with hot summers and mild winters will fare better starting tomatoes in late summer to grow throughout the fall and into winter.

Tomato seeds germinate more quickly if the potting medium is kept warm. After they are up, the seedlings grow fast indoors. When transplanting the plants outside, bury the stems to the top set of leaves. The underground portion of the stem will send out roots and create a stronger plant.

Stake the plants when you plant them and secure the branches as they grow. 'Green Grape' is not a large plant, but it will get heavy when it is full of fruit. It is susceptible to the same leaf and fruit diseases as other tomato plants. Keep the leaves dry, and do not plant tomatoes in the same location in consecutive seasons.

How to harvest Cherry tomatoes at the top of the cluster ripen first. 'Green Grape' is ready to pick when the blossom end of the fruit is a golden yellow.

Tip This is a relatively small plant with a large yield, making it perfect for containers.

Exposure:
Full sun

Ideal soil temperature:
75–85°F (24–30°C)

Planting depth:
1/4–1/2 in. (0.6–1 cm)

Spacing:
3–4 ft. (0.9–1.2 m)

Days to germination:
6–14 days

Days to maturity:
80–90 days

'Green Grape' is actually pale yellow when at its peak and orange when fully mature.

Others to try 'Current Gold Rush' is another small plant with huge yields; fruits are delicate, with a strong tomato flavor. 'Red Fig' produces small, juicy, pear-shaped tomatoes that dry nicely. 'Tommy Toes' produces 1 in. (2.5 cm) tangy fruits that hang in long clusters that resemble garlands.

'Opalka' is an amazing little package that manages to be juicy and sweet, and dense and meaty. When we think of tomatoes for making sauce, a lot of Italian names jump to mind. 'Roma' and 'San Marzano' may be familiar, but the Polish 'Opalka' is a welcome surprise. It is not often that a great paste tomato makes an equally great fresh eating tomato, although when you think about it, why not? The flavor has to be there or the sauce will not stand on its own.

'Opalka' is almost all flesh, with few seeds, making it a bit of a tease for seed-saving heirloom lovers. It was brought to the attention of gardeners by the godmother of heirloom tomatoes, Dr. Carolyn Male. It came to New York around 1900 with the Swidorski family and was given to Dr. Male by Carl Swidorski, a coworker whose wife's maiden name was Opalka.

Tomato 'Opalka'

Paste tomato
Lycopersicon lycopersicum

Flavor Texture plays a big part in the enjoyment of these tomatoes. Although juicy, they have a firm flesh with little watery gel. The skin is thin and tender, making it pleasant to bite into, and the flavor is rich but clean. It makes a luscious sauce and also slices well. Its dense texture makes it a favorite for sandwiches.

Growing notes 'Opalka' germinates easily and quickly. Start seed indoors 6–8 weeks before the expected transplant date. Move seedlings into individual 3–4 in. (7.5–10 cm) pots when true leaves appear. Transplant seedlings outdoors 2–4 weeks after your last frost. Wait until temperatures are reliably above 50°F (10°C) to set plants outside. Gardeners in areas with hot summers and mild winters will fare better by starting tomatoes in late summer to grow throughout the fall and into winter.

When transplanting the plants outside, bury the stems to the top set of leaves. The underground portion of the stem will send out roots and create a stronger plant.

Stake the plants when you plant them and secure the branches as they grow. 'Opalka' sets a heavy crop and needs a strong support. It has wispy foliage that can be prone to the leaf diseases, such as early blight and Septoria leaf spot. Keep the leaves dry and do not plant tomatoes in the same location in consecutive seasons.

How to harvest 'Opalka' is an indeterminate tomato, so the fruits do not ripen all at once, as do many paste tomatoes. 'Opalka' tends to ripen in clusters.

Others to try 'Federle' is drier than 'Opalka' but has a great flavor for making sauce. 'Green Sausage' produces sweet and tangy, yellow-green banana-shaped fruits. 'Purple Russian' tomatoes are blemish-free and fresh tasting.

Exposure:
Full sun

Ideal soil temperature:
75–85°F (24–30°C)

Planting depth:
1/4–1/2 in. (0.6–1 cm)

Spacing:
3–4 ft. (0.9–1.2 m)

Days to germination:
6–14 days

Days to maturity:
80–85 days

'Opalka' can hang onto large clusters of tomatoes, but it is prone to losing its leaves.

'Riesentraube' packs dense flavor into a cherry-sized tomato. The fruits may be small, but there are plenty of them, with each tall plant producing hundreds of blossoms during a season. When grown a little on the lean side, the complex flavors of this cherry tomato become concentrated and give it a character most often associated with much larger fruits.

A closet favorite among heirloom tomato lovers, 'Riesentraube' is finally making its way into the mainstream. The German name translates as giant grape, but it is actually the quantity of grapes in the clusters that is giant—often with 20 or more tomatoes per bunch. Food writer William Woys Weaver says the grape reference is for the biblical grapes of Eshcol, which promised a land of plenty. The 1 in. (2.5 cm) wide, plum-shaped, cherry tomatoes grow in long, heavy garlands and are easy to pick.

Tomato 'Riesentraube'

Cherry tomato
Lycopersicon esculentum

Flavor 'Reisentraube' is a fresh eating and snacking tomato with a rich flavor that is often likened to beefsteak tomatoes. The savory-sweet flavor is enhanced by a full aroma.

Growing notes Start seed indoors 6–8 weeks before expected transplant date. Move seedlings into individual 3–4 in. (7.5–10 cm) pots when true leaves appear. Transplant seedlings outdoors 2–4 weeks after your last expected frost. Wait until temperatures are reliably above 50°F (10°C). Gardeners in areas with hot summers and mild winters will fare better starting tomatoes in late summer to grow throughout the fall and into winter.

When transplanting the plants outside, bury the stems to the top set of leaves. The underground portion of the stem will send out roots and create a stronger plant.

The fruits may be small, but the sheer quantity of them means the plants need strong staking. The branches tend to shoot out in all directions, so caging may be the best option.

Although 'Reisentraube' is still susceptible to leaf and fruit diseases, it shows considerable resistance to most fungal problems. Keep the leaves dry and do not plant tomatoes in the same location in consecutive seasons.

How to harvest The large bunches of tomatoes ripen from the top down. Harvest those that turn bright red and are slightly soft. They will keep coming throughout the season.

Tip 'Riesentraube' can be used to make tomato wine. You can find several tomato wine recipes on the Internet. Their taste is described variously as similar to a dry white wine or a pale sherry. It should be fun to experiment.

Others to try 'Chadwick's Cherry' is a prolific producer with a balanced sweetness. It was developed by the late

Exposure:
Full sun

Ideal soil temperature:
75–85°F (24–30°C)

Planting depth:
1/4–1/2 in. (0.6–1 cm)

Spacing:
3–4 ft. (0.9–1.2 m)

Days to germination:
6–14 days

Days to maturity:
80–85 days

'Riesentraube' ripens slowly down its clustered stem of tomatoes for a lengthy snacking season.

horticultural guru and breeder Alan Chadwick. 'Grandpa's Minnesota' has good sugar content that keeps it sweet. 'Yellow Riesentraube' is similar though not as prolific.

This sunny yellow tomato is an intriguing heirloom, because so many people have memories of eating them at their grandmother's house. It seems to be a universal family heirloom. 'Yellow Pear' is deceptively vigorous. It is often the first tomato to set and ripen and the last to stop producing. Some gardeners complain that it produces too many tomatoes, so many fall to the ground and seed themselves. There are worse problems to have.

This old variety dates back to the 1700s. Despite its diminutive size, 'Yellow Pear' was grown for making tomato preserves or jam. Its light acidity and dense, juicy flesh make it great for preserves, plain or spicy.

Exposure:
Full sun

Ideal soil temperature:
75–85°F (24–30°C)

Planting depth:
1/4–1/2 in. (0.6–1 cm)

Spacing:
3–4 ft. (0.9–1.2 m)

Days to germination:
6–14 days

Days to maturity:
70–80 days

'Yellow Pear' is one of the most popular tomatoes sold at farmers' markets, but it is easy to grow in your own garden.

Tomato 'Yellow Pear'
Cherry tomato
Lycopersicon lycopersicum

Flavor 'Yellow Pear' is a subtle seducer, with a gentle, sweet flavor that teases just enough to encourage you to have one more. And one more. It is the perfect little garnish, makes a nice addition to any salad or antipasto plate, and tastes great in salsa.

Growing notes Start seed indoors 6–8 weeks before the transplant date. Move seedlings into individual 3–4 in. (7.5–10 cm) pots when true leaves appear. Transplant seedlings outdoors 2–4 weeks after your last expected frost. Wait until temperatures are reliably above 50°F (10°C). Gardeners in areas with hot summers and mild winters will fare better starting tomatoes in late summer to grow throughout the fall and into winter.

When transplanting the plants outside, bury the stems to the top set of leaves. The underground portion of the stem will send out roots and create a stronger plant. Go easy on the fertilizer to concentrate the flavor in the fruits.

'Yellow Pear' is a surprisingly large plant for a cherry tomato. Provide strong staking to keep the branches from drooping and touching the soil.

Although it is still susceptible to leaf and fruit diseases, 'Yellow Pear' usually survives until late in the season. Fruits tend to crack after a lot of rain.

How to harvest Fruits turn a vivid yellow when they are ripe. You will be able to smell the sugar, and the fruits will feel a bit soft when gently squeezed.

Others to try 'Blondkopfchen' (little blonde girl) is a golden cherry tomato from Germany, with a crisp, citrusy flavor. 'Red Pear' is similar to 'Yellow Pear', and when placed together in a bowl, they create delectable eye candy. 'Snowberry' is a pale yellow cherry with a sweet and fruity taste.

'Yellow Stuffer' is neither juicy nor sweet, but it can be used to create a delicious little bowl for stuffing. If you like baked tomatoes and do not want the disappointment of seeing them turn into sad, mushy bundles in the oven, you will love this heirloom.

Stuffing tomatoes resemble bell peppers more than slicing tomatoes. The origin of this tomato is not clear, but it probably resulted from a cross with 'Zapotec Pleated' tomato, an old variety grown in Oaxaca, Mexico, long before the Spanish arrived. The fruit is fluted or lobed and conveniently holds its seeds in a small cluster attached under the stem. Slice off the top to reveal little more than thick, fleshy walls and some ribs. There is too little demand for these to be grown commercially, so they remain an heirloom gardener's secret.

Tomato 'Yellow Stuffer'
Lycopersicon lycopersicum

Flavor The flavor of 'Yellow Stuffer' is not as acidic as we expect in yellow tomatoes. It has a subtle, fruity tomato flavor that does not overpower the stuffing. It probably will not win any contests for fresh eating, but sliced rings as a garnish are always welcome. It knows its place and complements the foods it carries, whether that be a baked and cheesy stuffing or a simple chilled salad.

Growing notes Start seed indoors, 6–8 weeks before the expected transplant date. Move seedlings into individual 3–4 in. (7.5–10 cm) pots when true leaves appear. Transplant seedlings after the weather stays reliably above 50°F (10°C), even at night. Gardeners in areas with hot summers and mild winters can start tomatoes in late summer to grow throughout the fall and into winter.

Tomato seeds will germinate more quickly if the potting medium is kept warm. Once they are up, the seedlings grow fast indoors. When transplanting the plants outside, bury the stems to the top set of leaves. The underground portion of the stem will send out roots and create a stronger plant.

Stake the plants when you plant them and secure the branches as they grow. 'Yellow Stuffer' is susceptible to the same leaf and fruit diseases as other tomato plants. Keep the leaves dry and do not plant tomatoes in the same location in consecutive seasons.

Allow the plants plenty of room for air circulation to prevent the spread of diseases, keep the branches up off the ground, and preserve the appearance of the fruits. Presentation is more important with stuffing tomatoes than with slicers because they are served whole.

How to harvest Harvest while the tomatoes are still glossy and a glowing yellow. Cut off the fruits to prevent injury to the branches.

Tip 'Yellow Stuffer' can produce more fruit than you care to

Exposure:
Full sun

Ideal soil temperature:
75–80°F (24–27°C)

Planting depth:
1/4–1/2 in. (0.6–1 cm)

Spacing:
3–4 ft. (0.9–1.2 m)

Days to germination:
6–14 days

Days to maturity:
80–85 days

Because of its firm shell, 'Yellow Stuffer' tends to keep longer than juicy tomatoes.

eat at one time. You can clean and freeze them to use later, and they will hold their shape and firmness.

Others to try 'Pink Accordion' is a flavorful slicer but shaped like a stuffer. 'Striped Cavern' is a small heirloom that can be served as an appetizer. 'Zapotec Pink Pleated' is sweet and meatier than most stuffers.

Enveloped in a papery husk, 'Purple' tomatillos hide out among the plant's foliage until they begin to flush and finally burst open to display their glowing purple cheeks. This pretty plant is finally getting some notice as more of us embrace spicier foods. 'Purple' tomatillo is too charming and delectable to relegate to an occasional treat, and tomatillos can be more than just salsa ingredients. The Aztecs knew this, and they grew and cultivated tomatillos as far back as 800 BC.

Exposure:
Full sun

Ideal soil temperature:
65-75°F (18-24°C)

Planting depth:
Barely cover with soil

Days to germination:
7-10 days

Spacing:
2-3 ft. (0.6-0.9 m)

Days to maturity:
90-100 days

'Purple' tomatillos turn purple when they are ready to be harvested.

Tomatillo 'Purple'

Mexican husk tomato
Physalis ixocarpa

Flavor 'Purple' tomatillos are smaller and sweeter than the traditional tart, green tomatillo often used in salsas. The spicy, sweet flavor reminds me of an herbal tomato. Many recipes call for raw tomatillos, but 'Purple' tomatillos mellow nicely when cooked and especially when roasted, which sweetens and intensifies their qualities. They also make great additions to soups and stews.

Growing notes Start seed indoors in a warm spot, 6–8 weeks before expected transplant date. Move seedlings into individual 3–4 in. (7.5–10 cm) pots when true leaves appear. Keep them indoors until the temperature stays around 50°F (10°C), even at night. Transplant seedlings outdoors 2–4 weeks after your last expected frost. Gardeners in warmer climates can direct seed in the ground. Tomatillo is a tropical perennial, hardy to Zone 10. Gardeners with warm winters can plant in midsummer to late summer and harvest into winter. Tomatillos have a tendency to self-seed profusely if fruits are left on the ground over winter.

Like peppers and tomatoes, tomatillos will root from the base of the stem, so transplant them a little deeper than they sit in their pots. If you are direct seeding, plant the seeds about 1/4 in. (0.6 cm) deep and then mound soil around the emerging stem.

Tomatillos need regular watering to get established, but they can handle drier conditions after that. Pests and disease are not usually a problem. Add mulch around the plants, because the fruits can sometimes droop and touch the ground.

Tomatillos are self-sterile, so at least two plants are required to produce fruit.

How to harvest A tomatillo is ripe when the husk breaks open and turns yellow or brown. Remove the husk; the tomatillo inside often has an oily feel to it, so wash it off before storing.

Others to try 'Dr. Wyche's Yellow' is similar in flavor to the green varieties but is yellow when ripe. 'Green Husk' is a sweet, green variety used for traditional salsas. 'Purple de Milpa' is a smaller, tarter, purple variety.

215

TOMATOES, TOMATILLOS, AND GROUND CHERRIES

There is no way to prepare yourself for the burst of flavor when you bite into an 'Aunt Molly's' ground cherry. Although ground cherries can be found growing wild, the cultivated varieties have a more refined taste. They are thought to have originated in Central or South America, but 'Aunt Molly's' comes from Poland. Polish growers have done an exemplary job of saving and passing along all kinds of seed. The first recorded reference to this ground cherry in the United States was in Pennsylvania in 1837. Unfortunately, we do not know much about Aunt Molly herself.

Exposure:
Full sun to partial shade

Ideal soil temperature:
75–80°F (24–27°C)

Planting depth:
1/4–1/2 in. (0.6–1 cm)

Spacing:
12–24 in. (30.5–61 cm)

Days to germination:
7–21 days

Days to maturity:
70 days

The fruits of 'Aunt Molly's' resemble small tomatillos, but their flavor is distinct.

Ground cherry 'Aunt Molly's'

Strawberry tomato, husk tomato
Physalis pruinosa

Flavor This fruit is so sweet it could be included in a fruit salad. You will note a little tomato flavor to be sure, but you can also taste flavors of pineapple, strawberry, citrus, and the slightest touch of vanilla. 'Aunt Molly's' ground cherry is a refreshing snack eaten fresh and it can also be used for baking, jams, and preserves. Try it with chocolate. The fruits need to be fully ripe before you eat them, however, or they can have an unpleasant flavor.

Growing notes Start seed indoors in a warm spot, 6–8 weeks before the expected transplant date. Move seedlings into individual 3–4 in. (7.5–10 cm) pots when true leaves appear. Keep them indoors until the temperature stays around 50°F (10°C), even at night. Transplant seedlings outdoors 2–4 weeks after your last expected frost. Gardeners in areas with hot summers and mild winters can direct sow seed in late summer to grow throughout the fall and into winter.

Seeds germinate quickly, and plants may self-seed for years

to come. Like its wild relatives, 'Aunt Molly's' is an undemanding plant. It grows quickly in the spring and should start flowering by early summer and fruiting in midsummer. Water will keep it growing, but ease back as the fruits start to form. Too much water will dilute the flavor and cause the fruits to crack.

To get a good yield, you will need several plants for pollination. 'Aunt Molly's' is happy to crawl along the ground, but it is also nice flopping over containers and hanging baskets. Guard the fruit as it ripens or the birds will beat you to it.

How to harvest Fruit turns a deep golden orange when it is ripe, and the husks turn brown and start to pull away from the fruit inside. Fruits fall off the plant as they ripen, hence the name ground cherry. You can store them in their husks for 2–3 months.

Tip Wait for these fruits to ripen fully; the unripe green fruits are toxic if ingested.

Others to try 'Cossack Pineapple' has pronounced pineapple flavor. 'Goldie' ripens later and is larger than most ground cherries. Peruvian ground cherry or cape gooseberry (*Physalis peruviana*) has a similar flavor.

WINTER SQUASH

It is a wonderful thing to go to the cupboard in late winter and find and cook vegetables harvested from your garden. Winter squash bides its time, preserved in hard shells of the most stunning colors, ranging from dusky rusts to deep blue-greens. Winter squash are grown to bake into side dishes that scent the whole house and lure you to the table on a chilly winter evening. The luscious aromatic flesh is rich and satisfying.

Winter squash is a part of the traditional Native American "three sisters" planting method. Squash is planted under corn to act as a mulch and to keep animals away from the stalks. The third sister, beans, helps feed the soil by fixing nitrogen in its roots, and it uses the corn as a trellis on which to climb.

Exposure:
Full sun

Ideal soil temperature:
70–75°F (21–24°C)

Planting depth:
1–1½ in. (2.5–4 cm)

Days to germination:
8–10 days

Spacing:
6–8 in. (15–20 cm)

Days to maturity:
85 days

'Baby Blue' squash is the perfect winter comfort food. After the exertion of slicing open a rock-hard shell, you're rewarded when it is cooked, with its sumptuous creamy, sweet, golden yellow puree that would make a pumpkin pie jealous.

Developed in 1953 at the University of New Hampshire, beautiful 'Baby Blue' is a young heirloom. This winter squash will store for months and provide many a hearty meal during the winter. You can feed the neighborhood with a full-sized Hubbard squash. The fruits can weigh as much as 50 lb. (23 kg). 'Baby Blue' is a much more manageable 5–6 lb. (2–3 kg), and each plant should produce four to six squash.

Squash 'Baby Blue'
Hubbard squash
Cucurbita maxima

The small, lacquered fruits of 'Baby Blue' Hubbard become heavy and weigh down the vines; if you use a trellis, make it a sturdy one.

Flavor Sweet and soul-warming, the fragrance of cooking Hubbard squash can fill the house with the aroma of autumn. The flesh is fine-grained, creamy, and

hearty. 'Baby Blue' is at its best when baked or boiled and mashed.

Growing notes Direct sow in the garden 2–3 weeks after the last expected frost. You can also start seed indoors, 3–4 weeks before the transplant date. 'Baby Blue' has a slightly shorter season than full-sized Hubbard squash and can be started in midsummer in frost-free climates to harvest in early fall.

Well-behaved in the garden, the plants grow in a bushy form , rather than sprawling like most squash plants. Keep the area free of weeds until the vines cover the ground. At that point, the squash can pretty much take care of itself.

If you want to trellis the plants, make sure the support is strong enough to hold at least 30 lbs. (14 kg) of fruit.

Squash bugs and cucumber beetles can do a lot of damage to all kinds of squash plants. Try to deal with the problem early by watching for their orange eggs on the undersides of leaves. Squash the eggs before they ruin your squash.

When to harvest 'Baby Hubbard' will signal it is ready for harvest when you cannot pierce the shell with your fingernail. Its color starts to dull as it matures, and the stem end may crack. Harvest all fruits before a frost and use the less mature fruits within a month.

Tip A less labor-intensive way to crack open a 'Baby Blue' Hubbard is to put it in a large plastic bag and drop it on the ground.

Others to try 'Baby Green' is similar in flavor and tastes great in pies or baked as a side dish. 'Gill's Sugar' is larger, with each squash about 25 lb. (11.5 kg), with sweet, pale orange flesh and a smooth shell. 'Little Gem' is an orange-skinned Hubbard with dryer flesh at 3–5 lb. (1–2 kg).

The shock of 'Lady Godiva' squash is more of a pleasant surprise than a bombshell: the seeds are hulless. If you love eating pumpkin seeds, you will be delighted to scoop up handfuls of them without being required to pop each one out of its shell. But there is more to this squash. The seeds are about three times the size of a hulled pumpkin seeds, with about 3–6 oz. (85–170 g) of seeds per fruit and about a dozen fruits per plant.

Although 'Lady Godiva' is in fact a pumpkin, it does not look like one. The oval shaped fruits are green with orange stripes and weight about 5 lb. (2.5 kg) each. 'Lady Godiva' also makes an exotic Jack-o-lantern.

Exposure:
Full sun

Ideal soil temperature:
70–75°F (21–24°C)

Planting depth:
1–1½ in. (2.5–4 cm)

Days to germination:
8–10 days

Spacing:
2–3 ft. (0.6–0.9 m)

Days to maturity:
80–100 days

Squash 'Lady Godiva'
Pumpkin
Cucurbita pepo

Flavor The green seeds are tender with a nutty flavor, even when eaten raw. Roasted, toasted, or sautéed, they make a wonderful addition to salads, stir-fries, and pastas, as well as a nutritious snack on their own. The seed contains 38 percent protein as well as other nutrients. They will swell and make a

'Lady Godiva' pumpkin is ready to harvest when it shows green and orange stripes

popping noise in the oven, but they do not actually explode.

The edible flesh can be stringy when cooked. Most gardeners grow the plants for the seeds alone.

Growing notes Direct sow seeds in the garden 2–3 weeks after the last expected frost. You can also start seed indoors, 3–4 weeks before the transplant date. Gardeners in warm climates can plant in midsummer for a fall harvest. Only gardeners in Zones 10 and warmer can overwinter 'Lady Godiva'.

Because seeds have no protective hull, they are vulnerable to rotting in cool, wet soils, so plant them shallowly. Coating them with a fungicide before planting is recommended. Plant 4–5 seeds in a mound and thin seedlings to the two strongest plants. You will need at least two vines to have enough flowers for adequate pollination and fruit production.

Keep the vines well watered. You can trellis plants or simply let them wander. If you are growing these squash for the seeds, misshapen fruits are not an issue. Watch for squash bugs and cucumber beetles.

How to harvest 'Lady Godiva' is ready to harvest when the skin starts to feel tough and hard. It should be green with orange stripes; it is past its prime when the skin starts to turn yellow. Be sure to harvest before a hard frost or the squash will rot.

Tip If you want to save seeds for planting, place 'Lady Godiva' away from other squash varieties to avoid cross-pollination; if this occurs, seeds from the next year's plants will not be hulless.

Others to try 'Triple Treat' and 'Hungarian Mammoth' both also have hulless seeds. 'Triple Treat' is similar to 'Lady Godiva'. *Cucurbita maxima* 'Hungarian Mammoth' has a harder shell and stores well.

'Potimarron' is a hearty winter squash with a delicate, refined flavor. Its name is a combination of the French words for pumpkin and chestnut, and it does have a depth of chestnut flavor when baked.

This beautiful fig-shaped squash has a deep, true orange color and is large enough to share. 'Potimarron' develops its full flavor as it matures, and it is a good keeper. If you cannot wait, it can be eaten as a summer squash while still immature and the skin is tender. The chestnut flavor is less intense when the squash is eaten while young, but it offers a dense, meaty texture and a more savory flavor.

Squash 'Potimarron'

Pumpkin
Cucurbita maxima

Flavor 'Potimarron' is undeniably rich, like a prohibited dessert. The aroma is a good part of its charm, and roasting or baking it lets you enjoy the flavor to its fullest. It can also be pureed or sautéed, especially when young and tender. It is so versatile that it can be used in place of either zucchini or pumpkin in recipes.

Growing notes Direct sow seeds in the garden 2–3 weeks after the last expected frost. You can also start seed indoors, 3–4 weeks before the transplant date. Gardeners in frost-free areas can sow seeds in early spring. Gardeners in Zones 10 and warmer can sow a second crop in midsummer.

This climber can be grown on a trellis, but it matures fast enough to sprawl on the ground without the fruits being damaged. Start with a rich soil, and keep the area free of weeds. The fruits start forming early in the season and mature before those of many other winter squash.

'Potimarron' is undemanding. After vines begin growing, supply water when the soil gets dry. You can feed the vines in midsummer with a balanced fertilizer, but if your soil is rich to begin with, supplemental feeding is not necessary. Watch for the usual squash pests, such as squash bugs and cucumber beetles, and ward off powdery mildew by making sure the vines are not crowded and have plenty of sunshine.

When to harvest 'Potimarron' can be harvested while it is still green. If you want the full chestnut flavor and a squash to store, allow it to turn a deep orange, but do not wait until the skin becomes hard and tough.

Tip The longer 'Potimarron' is stored, the better its sugar and vitamin content.

Others to try 'Golden Delicious' has a similar aroma and flavor in a larger size. 'Lakota' has creamy, nutty flesh and was once prized by the Lakota Sioux. 'Red Kuri' is a Japanese squash with a sweet, dry flesh that bakes wonderfully.

Exposure:
Full sun

Ideal soil temperature:
70–75°F (21–24°C)

Planting depth:
1–1 1/2 in. (2.5–4 cm)

Days to germination:
8–14 days

Spacing:
6–8 ft. (2–2.5 m)

Days to maturity:
85–95 days

'Potimarron' is a teardrop-shaped pumpkin with a shiny, hard shell.

Heady, aromatic, and rich, 'Small Sugar' pumpkins have been creating holiday memories and tantalizing diners for centuries. It has been mentioned in seed catalogs dating back to at least 1860, when it was also known as New England pie pumpkin. The colonial version of pumpkin pie was a hollowed out pumpkin filled with milk and spices and baked in the ashes of the fire. 'Small Sugar' pumpkin, at about 8–12 in. (20–30.5 cm) in diameter and 8–10 lb. (3.5–4.5 kg), is perfect for any type of a pie.

To the American colonists, the pumpkin meant sustenance. Among the seeds given them by Native Americans was this pumpkin. Because it is a good keeper, it was eaten throughout the winter. 'Small Sugar' is an older and better behaved variety of the popular 'Connecticut Field' pumpkin, a rambling vine grown commercially.

Exposure:
Full sun

Ideal soil temperature:
70-75°F (21-24°C)

Planting depth:
1-1½ in. (2.5-4 cm)

Days to germination:
10-14 days

Spacing:
4-6 ft. (1.2-2 m)

Days to maturity:
95-100 days

'Sweet Dumpling' can work for dessert or dinner.

Squash 'Small Sugar'
Pie pumpkin
Cucurbita pepo

Flavor As its name implies, 'Small Sugar' is naturally sweet, with a thick, fine-grained, almost creamy consistency and no strings. It makes an excellent pie, but it is also wonderful cubed and used in soups, stews, and pasta.

Growing notes Direct sow seeds in the garden 2–3 weeks after the last expected frost. You can also start seed indoors, 3–4 weeks before the transplant date. Pumpkins grow faster in warm climates, where they can be started in early summer to midsummer. Only gardeners in Zones 10 or warmer can reliably grow pumpkins in the fall and early winter.

Start with a rich soil. Plant four or five pumpkin seeds together in a mound and thin seedlings to the two strongest plants. 'Small Sugar' vines are fairly long and require a minimum spread area of 5–8 sq. ft. (0.5–0.7 sq. m). At least two vines are required

to produce enough flowers for adequate pollination. Keep the vines well watered and the area free of weeds. Eventually the vines will act as their own mulch.

Two potential problems are powdery mildew, which can severely stress the plants, and cucumber beetles and squash bugs, which carry disease and can mark the fruits.

How to harvest Harvest when pumpkins turn a rich orange and the skin feels tough. Your fingernail should barely dent the surface. Use a knife to sever the stem from the plant. Harvest before a hard frost or pumpkins can rot.

Tip Most squash, including pumpkins, store better if a few inches of the stem remains on the fruit.

Others to try 'Australian Butter' is slightly larger with tan skin and sweet orange flesh. 'Boston Marrow' is a great choice for areas with short summers. 'Long Island Cheese' was named for its resemblance to a wheel of cheese; its orange flesh tastes great when baked.

The rich, sweet flavor and cream-colored shell with green and orange stripes and speckles tells you that 'Sweet Dumpling' is related to 'Delicata' squash. 'Sweet Dumpling' is more robust in flavor and is shaped like a top-flattened pumpkin. When the squash is small, you do not need to remove the skin when cooking, because it is tender enough to eat. With older squash, the flesh can be scooped out easily after cooking.

This is a single-serving squash, maxing out at 1–2 lb. (0.5–1.0 kg). It stores well but yields only about eight fruits per plant, so storing may not be an issue. This squash was originally marketed as "vegetable gourd" because it makes such a beautiful fall decoration. 'Sweet Dumpling' seemed a more delectable name and now it has become a trendy upscale designer vegetable. Everything old is new again.

COLORFUL

Squash 'Sweet Dumpling'

Vegetable gourd
Cucurbita pepo

Exposure:
Full sun

Ideal soil temperature:
70–75°F (21–24°C)

Planting depth:
1–1½ in. (2.5–4 cm)

Days to germination:
8–14 days

Spacing:
2–5 ft. (0.6–1.5 m)

Days to maturity:
90–100 days

Flavor 'Sweet Dumpling' squash combines a starchy, almost grainy flavor with a strong suggestion of sweet potato, except fresher. The sweetness is not overpowering or candylike, but a gentle back note that can be manipulated by seasonings. Spice it up for dessert or stuff it with contrasting savories for a meal.

Growing notes Direct sow seeds in the garden 2–3 weeks after the last expected frost. You can also start seed indoors, 3–4 weeks before the transplant date. It can be a finicky germinator, so you might plant a few extra seeds for insurance. This warm-season crop is not generally started in the fall, except in gardens in Zones 10 and warmer.

'Sweet Dumpling' has a moderately enthusiastic vine, as

'Sweet Dumpling' can work for dessert or dinner.

squash plants go. A trellis can be helpful especially for keeping the fruits off the ground. Powdery mildew can be a problem, so make sure plants have plenty of air circulation and sunshine.

Squash bugs and cucumber beetles can do a lot of damage to all kinds of squash plants. Try to catch the problem early by watching out for their orange eggs on the undersides of leaves. Squash the eggs before they ruin your squash. After vines are growing, water when the soil gets dry. Otherwise, 'Sweet Dumpling' can pretty much take care of itself.

When to harvest 'Sweet Dumpling' is ready to harvest when the stem starts to crack. Squashes can range from the size of an orange to that of a cantaloupe, but they are cream-colored with green stripes when they are ready. The stripes will change to orange in storage.

Tip Insufficient pollination can cause young fruits to shrivel and drop shortly after forming. Squash blossoms are open for only a day, and they need to be pollinated several times to form fruits. Take matters into your own hands: pluck a couple of male blossoms and dust the female flowers with them.

Others to try 'Delicata' is an oblong squash that is similar in color and with a sweet potato flavor. 'Kabocha' is creamy and honey sweet. 'Thelma Sanders Sweet Potato' is a productive, single-serving acorn type with a shorter growing season.

CREATING YOUR OWN HEIRLOOMS

We are no longer seed-savers by necessity. As a result, many vegetable variet-ies have been lost along with their unique genetic identities. Seeds may not look like much, but they are direct links between the past and the future. To-day we have access to a vast quantity of heirloom vegetable varieties because someone took the time to save seed.

Saving your own seed makes you a plant breeder. By selecting the best plants from which to harvest seed, you will develop a variety that is adapted to your particular garden environment. It will still be a 'Lemon' cucumber or a rat's tail radish, but it might have picked up a disease resistance or a tolerance for acidic soil along the way. By growing and sharing your own seeds, you cre-ate your own heirlooms.

Heirloom veg-etables re-mind us that vegetables are seasonal and regional, varying from one garden to the next.

SEED-SAVING BASICS

Seed-saving is not difficult, but it is not as simple as collecting seeds from dried pods at the end of the season. To ensure some reliability and true-to-type seed, you need to take a few strategies into account.

Choose the Best Seeds to Save

Save seed from only the best vegetables. Choose the largest seeds from vegetables with superior traits: the sweetest melon, the fat onion with a slender neck, the pea that produced well into summer. You want the strongest seed that performed best in your garden. You will not get great plants from poor seed.

Prevent Cross-Pollination

Plants within families can cross-pollinate, and the resulting seed might not resemble either parent. Insects hop from one plant to the next indiscriminately, spreading pollen along the way. Some plants, such as spinach, are wind-pollinated, and pollen can be blown a mile (1.6 km) or more away. So the first major hurdle in saving seed is isolation. Keep in mind that nearby neighbors may also be growing plants that could cross-pollinate with yours, so isolation is not always dependable.

You can isolate flowers during the pollination process in three ways.

1. Use distance. Separate plants known to cross-pollinate by growing them at specified distances from one another (see "Vegetable Seed-Saving Guide"). The distances will vary with botanical families and can depend on whether a vegetable is wind- or insect-pollinated.
2. Use timing. You can ensure pure seed if you stagger planting times of similar crops, starting one early in the season and waiting to plant the other until the first is already in bloom.
3. Use barriers. Physical barriers, such as bagging individual flowers, adding row covers, or caging whole plants to exclude insects, will prevent cross-pollination. Plants that need insects to spread their pollen can be isolated on alternating days, by exposing one variety while another remains covered.

Sometimes it is easiest to pollinate the plants yourself, by dusting pollen from the male flower onto the stigma of the female. Plants will still need to be controlled to be sure of purity, however.

Clean, Dry, and Store Seeds Properly

Some seeds, such as peas and beans, are easy to collect. They dry in their pods and can be shelled and stored. Others, such as tomatoes and melons, are surrounded by a wet growth-inhibiting gel and must be processed before the seed is viable. These issues are addressed according to the specific plants.

For most plants, allow the flowers or fruit to mature fully before harvesting seed, even to the point of letting them dry on the plant. If the weather turns wet or a frost is forecast, bring the plants indoors to finish drying. Wet seed becomes less viable and can quickly rot. To dry indoors, pull or cut the plants and hang them in a warm, dry spot for a week or two. Seeds must be dried to the point of being brittle before you store them in airtight or sealed containers. Even a small amount of moisture in a seed can cause it to rot. When in doubt, store your seed in paper envelopes or bags to allow air to circulate. You can use silica gel or even food dehydrators to dry seed, but do not use your oven, because even the lowest setting is still too hot. Seeds contain the embryo of a future plant, and temperatures warmer than 96°F (36°C) can damage that embryo.

Thoroughly dry seed can be stored in a freezer to kill any insects or eggs that might be hiding inside the seed. If you do not want a freezer full of seeds, you can remove them after a few days and store them elsewhere.

Finally, do not forget to label and date seed containers.

VEGETABLE SEED-SAVING GUIDE

Amaranth

Cross-pollinates: Yes, with other amaranth

Isolation distance: 500 ft. (152 m) with a tall barrier in between

How to save seed: Seed matures from the bottom of the flower stalk up. As seed starts to mature, shake the flowers into a bag each day. Thoroughly dry the seeds before storing.

Artichoke

Cross-pollinates: Rarely; artichokes do not often grow true to type. They are usually propagated by division.

Isolation distance: 100 ft. (30 m) with a tall barrier in between

How to save seed: Cut flowers when fully open. Place the flowers in a muslin bag and pound the bag with a hammer to dislodge the seeds.

Arugula

Cross-pollinates: Yes, with other varieties of arugula

Isolation distance: 1/2 mi. (0.8 km)

How to save seed: Collect dried seed pods and crush them in a bag. Winnow to separate the seed.

Asparagus

Cross-pollinates: Yes, with other asparagus

Isolation distance: 2 mi. (3.2 km)

How to save seed: Collect the red berries before they drop. Free the seeds by rubbing the berries on screening. Wash well and dry thoroughly before storing.

Bean

Cross-pollinates: Rarely

Isolation distance: Not necessary, but it is recommend that you not plant two white-seeded varieties next to each other, because you will not be able to discern whether they have crossed.

How to save seed: Allow pods to dry on the plants. Be sure the plants are disease-free. Harvest and shell by hand or crush and winnow large amounts.

Beet

Cross-pollinates: With other beets, as well as Swiss chard (silverbeet)

Isolation distance: 1/2 mi. (0.8 km)

How to save seed: Beets are biennials. During their second year they will bolt, and seed can be collected as the pods dry over a staggered period. Pods might need to be shielded from marauding birds. In areas where the ground freezes, beet roots can be dug and stored in damp peat or sawdust at 32–40°F (0–5°C), with 2–3 in. (5–7.5 cm) of leaves attached. Replant the bulbs in the spring.

Bok choy

Cross-pollinates: Yes, with all *Brassica rapa*

Isolation distance: 1 mi. (1.6 km)

How to save seed: Bok choy is a biennial, although it may bolt the first year. Unharvested plants can be dug, potted, and stored at 32–40°F (0–5°C). Replant in the spring, allow to flower, and harvest seed when the pods have dried.

Broccoli
Cross-pollinates: Yes, with all Brassica oleracea
Isolation distance: 1 mi. (1.6 km)
How to save seed: Broccoli is a biennial, although it may bolt the first year. Unharvested plants can be dug, potted, and stored at 32–40°F (0–5°C). Replant in the spring, allow to flower, and harvest seed when the pods have dried.

Broccoli raab
Cross-pollinates: Yes, with other Brassica rapa including broccoli raab, mizuna, turnips, and bok choy
Isolation distance: 1 mi. (1.6 km)
How to save seed: Some varieties will go to seed the first year; others can be left in the ground to send up stalks the following spring. Allow the pods to mature and thoroughly dry on the plant. Collect seeds before the pods break open.

Cabbage
Cross-pollinates: Yes, with all Brassica oleracea
Isolation distance: 1 mi. (1.6 km)
How to save seed: Cabbage is a biennial. Unharvested plants can be dug, potted, and stored at 32–40°F (0–5°C). Replant in the spring, allow to flower, and harvest seed when the pods have dried.

Carrots
Cross-pollinates: Yes, also with Queen Anne's lace, a wild carrot
Isolation distance: 1/2 mi. (0.8 km)
How to save seed: Carrots are biennials. Dig plants in the fall. Store with 1 in. (2.5 cm) of the tops attached in damp peat or sawdust, at 32–40°F (0–5°C). Replant in the spring,

Cauliflower
Cross-pollinates: Yes, with all Brassica oleracea
Isolation distance: 1 mi. (1.6 km)
How to save seed: Cauliflower is a biennial. Unharvested plants can be dug, potted, and stored at 32–40°F (0–5°C). Replant in the spring, allow to flower, and harvest seed when the pods have dried.

Celery and celeriac
Cross-pollinates: Yes, with other celery and celeriac
Isolation distance: 1 mi. (1.6 km)
How to save seed: Celery and celeriac are biennial plants. In areas with severe winters, lift plants in the fall and trim stalks to 2–3 in. (5–7.5 cm). Store in damp sand with the crowns exposed. Replant in the spring, allow to flower, and harvest seed when the pods have dried.

Collards
Cross-pollinates: Yes, with all Brassica oleracea
Isolation distance: 1 mi. (1.6 km)
How to save seed: Collards are biennials. Unharvested plants can be dug, potted, and stored at 32–40°F (0–5°C). Not all plants will survive, but those that do can be replanted in the spring, allowed to flower, and seed harvested when the pods have dried.

Corn
Cross-pollinates: Yes, with all corn varieties
Isolation distance: 1 mi. (1.6 km)
How to save seed: Allow ears to dry fully on the plants. If the growing season is too short, harvest ears when the husks are brown, pull

back the husks, and store them until kernels are completely dry. Harvest and remove kernels from the cob by grasping and twisting the ear.

Cucumber

Cross-pollinates: Yes, with other cucumbers in the same species
Isolation distance: 1/4 mi. (0.4 km)
How to save seed: Allow fruits to ripen and turn yellow and soft. Cut lengthwise and scoop out the seed. Wash off pulp and allow it to dry thoroughly. Seeds will be brittle and will break, rather than bend, when dry.

Eggplant

Cross-pollinates: Yes, with other eggplant
Isolation distance: 1/4 mi. (0.4 km)
How to save seed: Allow fruits to ripen fully and become soft. Eggplant seeds are tiny and difficult to collect. Try grating the flesh and then place it in a bucket of water. Squeeze the pulp to dislodge the seeds. Seeds will sink to the bottom and the pulp will float to the top.

Fennel

Cross-pollinates: Yes, with other fennel
Isolation distance: 1/2 mi. (0.8 km)
How to save seed: Fennel plants are hardy biennials. Some will go to seed in their first year if planted early. Others can be left in the ground and will bolt to seed the following spring. Harvest seed when the pods have dried.

Garlic

Cross-pollinates: Garlic rarely sets seed; it is grown from cloves
Isolation distance: Not applicable
How to save seed: Harvest bulbs when the tops of the plants start to yellow and bend. Allow bulbs to cure for 1–2 weeks. Store in a dark, dry place.

Ground cherry

Cross-pollinates: Possibly
Isolation distance: 100 ft. (30 m)
How to save seed: Allow fruits to ripen fully. Squeeze seed pulp into a jar and fill with water. Shake vigorously to separate the seed, which should sink to the bottom. Drain off pulp, strain the seed, and allow it to dry thoroughly before storing.

Jerusalem artichoke

Cross-pollinates: Rarely. Seed produced is usually sterile. Jerusalem artichokes are generally produced from tubers.
Isolation distance: Not applicable
How to save seed: Dig and divide the tubers.

Kale

Cross-pollinates: Yes, with all *Brassica oleracea*
Isolation distance: 1 mi. (1.6 km)
How to save seed: Kale is a biennial. Unharvested plants can be dug, potted, and stored at 32–40°F (0–5°C). Replant in the spring, allow to flower, and harvest seed when the pods have dried.

Kohlrabi

Cross-pollinates: Yes, with all *Brassica oleracea*
Isolation distance: 1 mi. (1.6 km)
How to save seed: Kohlrabi is a biennial. In frost-prone areas, dig plants after a hard frost and remove leaves. Clip roots to 4–6 in. (10–15 cm) and store in damp sand. Replant in the spring, allow to flower, and harvest seed when the pods have dried.

Leek

Cross-pollinates: Yes
Isolation distance: 1 mi. (1.6 km)
How to save seed: Leeks are biennial. Choose healthy plants and store at 32–40°F (0–5°C) over winter. Replant in the spring, allow to flower, and harvest seed when the pods have dried.

Lettuce

Cross-pollinates: Rarely
Isolation distance: 25 ft. (7.5 m)
How to save seed: Allow plants to bolt and send up flower stalks. Collect seed as the pods dry, which may continue over a staggered period. Pods might need to be shielded from birds.

Lima bean

Cross-pollinates: Yes, with lima beans, but not with other beans
Isolation distance: 1 mi. (1.6 km)
How to save seed: Allow pods to dry on the plants. Be sure the plants are disease-free. Harvest and shell by hand or crush and winnow large amounts.

Melon

Cross-pollinates: Yes, with other melons
Isolation distance: 1/4 mi. (0.4 km)
How to save seed: Allow fruits to ripen fully, cut them open, and scoop out the seed. Wash off the pulp and allow seeds to dry thoroughly. Seeds will be brittle and will break, rather than bend, when dry.

Mizuna

Cross-pollinates: Yes, with other *Brassica rapa* including broccoli raab, mizuna, turnips, and bok choy
Isolation distance: 1 mi. (1.6 km)
How to save seed: Some varieties will go to seed the first year; others can remain in the ground to send up stalks the following spring.

Allow the pods to mature and thoroughly dry on the plant. Collect seeds before the pods break open.

Okra

Cross-pollinates: Yes, with other okra
Isolation distance: 1 mi. (1.6 km)
How to save seed: Allow pods to mature and turn brown on the plants. Harvest when they are just starting to split open, before seeds are scattered on the ground.

Onion

Cross-pollinates: Yes, with other onions
Isolation distance: 1 mi. (1.6 km)
How to save seed: Onions are biennial, although they may bolt the first year. Unharvested plants can be dug and stored at 32–40°F (0–5°C). Replant in the spring and allow to flower. When the flowers start to dry, cut them and place them in a paper or muslin bag to finish drying. Some seed will fall on its own. The rest can be removed by crushing the flower heads.

Parsnip

Cross-pollinates: Yes, with other parsnip varieties
Isolation distance: 1 mi. (1.6 km)
How to save seed: Parsnips are biennial. They should survive in the ground in all areas. Optionally, dig plants in the fall. Store with 1 in. (2.5 cm) of the tops attached, in damp peat or sawdust, at 32–40°F (0–5°C). Replant in the spring, allow to flower, and harvest seeds when the pods have dried.

Pea

Cross-pollinates: Yes, with other peas
Isolation distance: 50 ft. (15 m)
How to save seed: Allow pods to dry thoroughly on the plants. Harvest and shell by hand.

Pepper

Cross-pollinates: Yes, with other peppers

Isolation distance: 500 ft. (152 m)

How to save seed: Allow peppers to reach their mature color. Cut open and remove the seeds from the core. Allow to dry fully before storing.

Potato

Potatoes will sometimes produce seed pods, but they do not grow true to type from these seeds. Potatoes are grown from pieces of the potato tuber. Save only disease-free tubers for replanting.

Radish

Cross-pollinates: Yes, with other radishes

Isolation distance: 1/2 mi. (0.8 km)

How to save seed: Allow plants to flower. Harvest pods when they are brown and dry and remove seed by hand.

Rhubarb

Rhubarb does not grow true to type from seed. Plants are generally started from cuttings or root divisions.

Rutabaga

Cross-pollinates: Yes, with other rutabagas

Isolation distance: 1 mi. (1.6 km)

How to save seed: Rutabagas are biennial. They should survive in the ground in all areas, with mulching. Optionally, dig plants in the fall. Store with 1 in. (2.5 cm) of the tops attached, in damp peat or sawdust at 32–40°F (0–5°C). Replant in the spring, allow to flower, and harvest seed when the pods have dried.

Salsify

Cross-pollinates: Yes, with other salsify

Isolation distance: 1/2 mi (0.8 km)

How to save seed: Salsify is biennial. Dig plants in the fall. Store with 1 in. (2.5 cm) of the tops attached, in damp peat or sawdust at 32–40°F (0–5°C). Replant in the spring, allow to flower, and harvest seed when the pods have dried.

Sorrel

Cross-pollinates: Yes, with other sorrel of the same species

Isolation distance: 450 ft. (137 m)

How to save seed: Sorrel usually begins to flower in its second year. Allow seed to dry thoroughly on the plant before collecting. Sorrel is easily grown from plant divisions.

Soybean

Cross-pollinates: Rarely

Isolation distance: 100 ft. (30 m)

How to save seed: Allow pods to dry on the plants. Be sure the plants are disease-free. Harvest and shell by hand or crush and winnow large amounts.

Spinach

Cross-pollinates: Yes, with other spinaches

Isolation distance: 600 ft. (183 m)

How to save seed: Allow plants to bolt and send up flower stalks. Collect seed as the pods dry, which may occur over a staggered period. Seeds can be prickly and gloves may be needed.

Squash

Cross-pollinates: Yes, with squash in the same species

Isolation distance: 1/4 mi. (0.4 km)

How to save seed: Allow fruits to overripen by 3–4 weeks. Cut open and remove the seeds. Wash off the pulp and allow seeds to dry thoroughly before storing.

Sweet potato

Sweet potatoes are grown from rooted shoots called slips. You can create slips by cutting a clean sweet potato into large pieces. Place each piece in a glass and fill with water until the potato piece is covered halfway. You can use toothpicks to hold the potato in place. Place the glass in a warm spot and refill with water as needed. Leaves should sprout in 2 or 3 weeks. Twist each sprout off the potato piece and place it in a shallow pan of water. When roots emerge and reach 1–2 in. (2.5–5 cm) in length, the sprouts are considered slips and are ready to plant.

Swiss chard, silverbeet

Cross-pollinates: Yes, with silverbeets and regular beets
Isolation distance: 1/2 mi. (0.8 km)
How to save seed: Swiss chard is a biennial. During its second year it will bolt and seed can be collected as the pods dry, which may occur over a staggered period. Pods might need to be shielded from birds. In areas where the ground freezes, Swiss chard can be dug and stored in damp peat or sawdust at 32–40°F (0–5°C), with 2–3 in. (5–7.5 cm) of leaves attached. Replant in the spring.

Tomatillo

Cross-pollinates: No
Isolation distance: None required
How to save seed: Allow fruits to ripen fully. Squeeze seeds into a strainer and wash off the pulp. Allow seeds to dry thoroughly before storing.

Tomato

Cross-pollinates: Yes, with other tomatoes, especially heirlooms and other open-pollinated varieties
Isolation distance: 35 ft. (10.5 m) with a tall crop in between
How to save seed: Squeeze seed pulp from ripe fruits. Tomato seeds are surrounded by a gel that inhibits germination. To ensure the best viability, this gel will need to be fermented and removed. This process produces a strong odor and is best done outdoors. Place the pulp in an open jar and allow it to ferment for three days; stir it up twice per day. A gray mold will form on top. When the mold starts to bubble or completely covers the surface, add water to cover, close the jar and shake vigorously. Allow the seeds to settle to the bottom and pour off the mold. Repeat as necessary until the gel is removed, and then strain the liquid. Allow the seeds to dry thoroughly before storing.

Turnip

Cross-pollinates: Yes, with other *Brassica rapa* including broccoli raab, mizuna, turnips, and bok choy
Isolation distance: 1 mi. (1.6 km)
How to save seed: Some varieties will go to seed the first year; others can be left in the ground to send up stalks the following spring. Allow the pods to mature and thoroughly dry on the plant. Collect seeds before the pods break open.

Watermelon

Cross-pollinates: Yes, with other watermelons
Isolation distance: 1/4 mi. (0.4 km)
How to save seed: Allow fruits to ripen fully, remove seeds, and allow to dry thoroughly before storing.

GLOSSARY

Annual A plant that completes its entire life cycle in 1 year.

Biennial A plant that completes its life cycle in 2 years, growing foliage the first year and going to seed in the second.

Bottom heat To warm a seed-starting container from the bottom with the use of a heating mat or by placing the container on top of a warm appliance.

Broadcast To sow seed by tossing it onto an expanse of ground, rather than sowing it in a measured row.

Companion plants Plants that provide benefits to one another when planted in close proximity. Benefits can include pest resistance, physical support, or growth inducement.

Cure To prepare certain root crops for long-term storage by allowing them to dry naturally after being lifted from the ground.

Day length The response by plants to the length of sunlight and darkness. This response is called photoperiodism, and it can cause a plant to bloom or induce other responses, such as bulbing in onions.

Direct seed, direct sow To sow the seed directly in the garden, as opposed to starting it in pots and transplanting.

Ephemeral A plant that can die to the ground during hot weather. The roots remain viable and the plant resumes growing the following spring.

Harden off To acclimate seedlings grown indoors to outside conditions gradually by exposing them to the outdoors for increasing amounts of time until they are able to remain outside completely.

Keeper Slang term for a vegetable that can be stored for several months.

Perennial A plant that lives for 3 or more years.

Scoville scale A measurement of the amount of capsaicin in a pepper and the resulting intensity of heat. At the bottom of the scale are bell peppers with a measurement of zero (0). Pure capsaicin, the compound in peppers that make them hot, measures 16,000,000. All other peppers fall somewhere in between, such as jalapeno at 2500–8000; cayenne at 30,000–50,000; and habanero at 100,000–350,000.

Seedling A small plant started from seed in a pot.

Self-fertile A plant that can pollinate itself with its own pollen and produce seed. This is called self-pollination. The opposite of self-fertile is self-sterile, which refers to a plant that requires pollen from another plant to produce seed.

Transplant To plant a pot-grown seedling into the garden.

True leaves The first set of leaves to form after the initial seed leaves, or cotyledons.

Zone Short for hardiness zone. A zone rating refers to an area's average lowest annual temperature and provides an indication of whether a plant will survive the winter in that area.

HARDINESS ZONES

Temperatures
°C = 5/9 · (°F–32)
°F = (9/5 · °C) + 32

Plant Hardiness Zones
Average Annual Minimum Temperature

Zone	Temperature (deg. F)			Temperature (deg. C)		
1	BELOW		–50	–45.6	AND	BELOW
2A	–45	TO	–50	–42.8	TO	–45.5
2B	–40	TO	–45	–40.0	TO	–42.7
3A	–35	TO	–40	–37.3	TO	–40.0
3B	–30	TO	–35	–34.5	TO	–37.2
4A	–25	TO	–30	–31.7	TO	–34.4
4B	–20	TO	–25	–28.9	TO	–31.6
5A	–15	TO	–20	–26.2	TO	–28.8
5B	–10	TO	–15	–23.4	TO	–26.1
6A	–5	TO	–10	–20.6	TO	–23.3
6B	0	TO	–5	–17.8	TO	–20.5
7A	5	TO	0	–15.0	TO	–17.7
7B	10	TO	5	–12.3	TO	–15.0
8A	15	TO	10	–9.5	TO	–12.2
8B	20	TO	15	–6.7	TO	–9.4
9A	25	TO	20	–3.9	TO	–6.6
9B	30	TO	25	–1.2	TO	–3.8
10A	35	TO	30	1.6	TO	–1.1
10B	40	TO	35	4.4	TO	1.7
11	40	AND	ABOVE	4.5	AND	ABOVE

To see the U.S. Department of Agriculture Hardiness Zone Map, go to the U.S. National Arboretum site at www.usna.usda.gov/Hardzone/ushzmap.html.

RESORCES

Seed Sources: North America

Abundant Life Seed Company
P.O. Box 279
Cottage Grove, Oregon 97424
541-767-9606
www.abundantlifeseeds.com

Baker Creek Heirloom Seeds
2278 Baker Creek Road
Mansfield, Missouri 65704
417-924-8917
rareseeds.com

Colonial Williamsburg: The Colonial Nursery Seed List
The Colonial Nursery
P.O. Box 1776
Williamsburg, Virginia 23187-1776
757-229-1000
www.history.org/history/CWLand/nursery1.cfm

D. Landreth Seed Company
60 East High Street, Bldg #4
New Freedom, Pennsylvania 17349
800-654-2407
www.landrethseeds.com

Fedco Seeds
P.O. Box 520
Waterville, Maine 04903
207-873-7333
www.fedcoseeds.com

Filaree Garlic Farm
182 Conconully Highway
Okanogan, Washington 98840
509-422-6940
www.filareefarm.com

Heirloom Seeds
287 East Finley Drive
West Finley, Pennsylvania 15377
www.heirloomseeds.com

Heirloom Tomatoes
5423 Princess Drive
Rosedale, Maryland 21237
www.heirloomtomatoes.net

Heritage Harvest Seed
Box 40, RR3
Carman, Manitoba ROG Jo Canada
204-745-6489
heritageharvestseed.com

Hudson Valley Seed Library
www.seedlibrary.org

Johnny's Select Seeds
955 Benton Avenue
Winslow, Maine 04901
877-564-6697
www.johnnyseeds.com

Kitazawa Seed Company
P.O. Box 13220
Oakland, California 94661-3220
510-595-1188
www.kitazawaseed.com

Native Seeds/Southwestern Endangered Aridland Resource Clearing House
3061 North Campbell Avenue
Tucson, Arizona 85719
520-622-5561
www.nativeseeds.org

Old Sturbridge Village Seed Store
1 Old Sturbridge Village Road
Sturbridge, Massachusetts 01566
508-347-3362
www.osv.org

Plimoth Plantation
Plimoth Plantation Museum Shops
137 Warren Avenue
Plymouth, Massachusetts 02360
508-746-1622
www.plimoth.com

Sand Hill Preservation Center
1878 230th Street
Calamus, Iowa 52729
563-246-2299
www.sandhillpreservation.com

Seed Savers Exchange
3094 North Winn Road
Decorah, Iowa 52101
563-382-5990
seedsavers.org

Seeds of Change
P.O. Box 4908
Rancho Dominguez, California 90220
888-762-7333
www.seedsofchange.com

Seeds of Italy
P.O. Box 149
Winchester, Massachusetts 01890
781-721-5904
www.growitalian.com

Skyfire Garden Seeds
1313 23rd Road
Kanopolis, Kansas 67454
www.skyfiregardenseeds.com

**South Carolina Foundation
Seed Association**
Heirloom Collection,
Dr. David Bradshaw
1162 Cherry Road
Box 349952
Clemson, South Carolina 29634
864-656-2520
virtual.clemson.edu/groups/seed/
    heirloom.htm

Southern Exposure Seed Exchange
P.O. Box 460
Mineral, Virginia 23117
540-894-9480
www.southernexposure.com/index.html

**The Thomas Jefferson Center
for Historic Plants**
Monticello
P.O. Box 316
Charlottesville, Virginia 22902
800-243-1743
www.monticello.org

Vermont Bean Seed Company
334 West Stroud Street
Randolph, Wisconsin 53956
800-349-1071
www.vermontbean.com

Victory Seeds
P.O. Box 192
Molalla, Oregon 97038
503-829-3126
www.victoryseeds.com

Seed-Saving Networks: Europe

Irish Seed Savers Association
Capparoe, Scarriff
County Clare, Ireland
061 921866
www.irishseedsavers.ie

Arche Noah
The Austrian Seed Savers
Association
Obere Strasse 40
A-3553 Schiltern, Austria
43-(0)2734-8626
www.arche-noah.at/etomite/index.
    php?id=52

**Henry Doubleday Research
Association**
Heritage Seed Library
Garden Organic
Coventry, Warwickshire
CV8 3LG United Kingdom
44 (0) 24 7630 3517

SUGGESTED READING

Ashworth, Suzanne, and Kent Whealy. 2002. *Seed to Seed: Seed Saving and Growing Techniques for the Vegetable Gardener*. Decorah, Iowa: Seed Savers Exchange.

Baron, Robert C., and Thomas Jefferson. 1987. *The Garden and Farm Books of Thomas Jefferson*. Golden, Colorado: Fulcrum Publishing.

Burr, Fearing Jr. 2008. *Garden Vegetables*. Carlisle, Massachusetts: Applewood Books.

Deppe, Carol. 2000. *Breed Your Own Vegetable Varieties: The Gardener's and Farmer's Guide to Plant Breeding and Seed Saving*. White River Junction, Vermont: Chelsea Green Publishing.

Goldman, Amy. 2008. *The Heirloom Tomato from Garden to Table: Recipes, Portraits, and History of the World's Most Beautiful Fruit*. New York: Bloomsbury USA.

Goldman, Amy, and Victor Schrager. 2002. *Melons for the Passionate Grower*. New York: Artisan Publishers.

Goldman, Amy, and Victor Schrager. 2004. *The Compleat Squash: A Passionate Grower's Guide to Pumpkins, Squashes, and Gourds*. New York: Artisan Publishers.

Male, Carolyn J. 1999. *Smith & Hawken: 100 Heirloom Tomatoes for the American Garden*. New York: Workman Publishing Company Inc.

McLaughlin, Chris. 2010. *The Complete Idiot's Guide to Heirloom Vegetables*. New York: Alpha Books.

Mendelson, Kathy. 2008. *The Heirloom Vegetable Gardener's Assistant*. www.halcyon.com/tmend/heirloom.htm. Accessed 3 March, 2011.

Rogers, Marc, and Polly Alexander. 1990. *Saving Seeds: The Gardener's Guide to Growing and Storing Vegetable and Flower Seeds*. North Adams, Massachusetts: Storey Publishing.

Rowe, Jack. *Vegetable Seed Saving Handbook*. howtosaveseeds.com/seedsavingdetails.php. Accessed 3 March, 2011.

Schneider, Elizabeth. 2001. *The Essential Reference: Vegetables from Amaranth to Zucchini*. New York: HarperCollins Publishers.

Stickland, Sue, Kent Whealy, and David Cavagnaro. 1998. *Heirloom Veg-*

etables: *A Home Gardener's Guide to Finding and Growing Vegetables from the Past*. New York: Simon & Schuster.

Thorness, Bill. 2009. Edible Heirlooms: *Heritage Vegetables for the Maritime Garden*. Seattle: Mountaineers Books.

Toussaint-Samat, Maguelonne. 2009. *A History of Food 2nd ed. West Sussex, England*: Blackwell Publishing, Ltd.

Vilmorin-Andrieux, M. 1981. *The Vegetable Garden*. Berkeley, California: Ten Speed Press.

Watson, Benjamin. 1996. *Taylor's Guide to Heirloom Vegetables: A Complete Guide to the Best Historic and Ethnic Varieties*. Boston: Houghton Mifflin Company.

Woys Weaver, William. 1997. *Heirloom Vegetable Gardening: A Master Gardener's Guide to Planting, Seed Saving, and Cultural History*. New York: Henry Holt and Company, Inc.

Woys Weaver, William, and Signe Sundberg-Hall. 2000. *100 Vegetables and Where They Came From*. Chapel Hill, North Carolina: Algonquin Books of Chapel Hill.

INDEX

244

PHOTOGRAPHY CREDITS

Photographs are by the author except for the following:

About the Author

Longtime Master Gardener Marie Iannotti is the former owner of Yore Vegetables, an heirloom seedling nursery. She has served as a Cornell Cooperative Extension Horticulture educator and as a Master Gardener program coordinator and is a member of the Garden Writers Association and The Garden Conservancy.

Marie's garden writing has been featured in newspapers and magazines throughout the United States, and she has been interviewed for Martha Stewart Radio, National Public Radio, and numerous articles. Marie's blog is practicallygardening.com, and she is the gardening guide for About.com. She is the former editor of The Mid-Hudson Gardener's Guide.

She has gardened from coast to coast in the United States, growing thousands of varieties of plants in her own gardens and helping many beginners become confident, successful gardeners.